T0229849

Machine Learning for Mobile Communications

Machine Learning for Mobile Communications will take readers on a journey from basic to advanced knowledge about mobile communications and machine learning. For learners at the basic level, this book volume discusses a wide range of mobile communications topics from the system level, such as system design and optimization, to the user level, such as power control and resource allocation. The authors also review state-of-the-art machine learning, one of the biggest emerging trends in both academia and industry. For learners at the advanced level, this book discusses solutions for long-term problems with future mobile communications such as resource allocation, security, power control, and spectral efficiency. The book brings together some of the top mobile communications and machine learning experts throughout the world, who contributed their knowledge and experience regarding system design and optimization.

This book:

- Discusses the 5G new radio system design and architecture as specified in 3GPP documents.
- Highlights the challenges including security and privacy, energy, and spectrum efficiency from the perspective of 5G new radio systems.
- Identifies both theoretical and practical problems that can occur in mobile communication systems.
- Covers machine learning techniques such as autoencoder and Q-learning in a comprehensive manner.
- Explores how to apply machine learning techniques to mobile systems to solve modern problems.

This book is for senior undergraduate and graduate students and academic researchers in the fields of electrical engineering, electronics and communication engineering, and computer engineering.

Industry 5.0: Artificial Intelligence, Cyber Physical Systems, Wireless Communications and Smart Grids
Editor: Kolla Bhanu Prakash

This book series is aimed at discussing the fundamentals, advanced concepts and applications of important fields of research including artificial intelligence, cyber physical systems, wireless communication, mechatronics and smart grids. It will cover emerging trends in Ad hoc networks, cyber forensic, human augmentation, quantum cryptography, quantum machine learning, Hyperledger programming, remote network acquisition tools, blockchain dynamics, and nature inspired algorithms. Aimed at senior undergraduate students, graduate students and professionals, the proposed series will focus on the advances in industry 5.0, mechatronics, smart grids, artificial intelligence and cyber physical systems.

Machine Learning for Mobile Communications
Edited by Sinh Cong Lam, Chiranji Lal Chowdhary, Tushar Hrishikesh Jaware and Subrata Chowdhury

For more information about this series, please visit: www.routledge.com/Industry-50/book-series/CRCINDUSTRY5

Machine Learning for Mobile Communications

Edited by
Sinh Cong Lam, Chiranji Lal Chowdhary,
Tushar Hrishikesh Jaware, and
Subrata Chowdhury

CRC Press
Taylor & Francis Group
Boca Raton London New York

CRC Press is an imprint of the
Taylor & Francis Group, an **informa** business

Designed cover image: Sinh Cong Lam, Chiranji Lal Chowdhary,
Tushar Hrishikesh Jaware and Subrata Chowdhury

First edition published 2024
by CRC Press
2385 NW Executive Center Drive, Suite 320, Boca Raton FL 33431

and by CRC Press
4 Park Square, Milton Park, Abingdon, Oxon, OX14 4RN

CRC Press is an imprint of Taylor & Francis Group, LLC

© 2024 selection and editorial matter, Sinh Cong Lam, Chiranji Lal Chowdhary,
Tushar Hrishikesh Jaware and Subrata Chowdhury; individual chapters, the
contributors

Reasonable efforts have been made to publish reliable data and information,
but the author and publisher cannot assume responsibility for the validity of
all materials or the consequences of their use. The authors and publishers
have attempted to trace the copyright holders of all material reproduced in this
publication and apologize to copyright holders if permission to publish in this
form has not been obtained. If any copyright material has not been acknowledged
please write and let us know so we may rectify in any future reprint.

Except as permitted under U.S. Copyright Law, no part of this book may be
reprinted, reproduced, transmitted, or utilized in any form by any electronic,
mechanical, or other means, now known or hereafter invented, including
photocopying, microfilming, and recording, or in any information storage or
retrieval system, without written permission from the publishers.

For permission to photocopy or use material electronically from this work, access
www.copyright.com or contact the Copyright Clearance Center, Inc. (CCC), 222
Rosewood Drive, Danvers, MA 01923, 978–750–8400. For works that are not
available on CCC please contact mpkbookspermissions@tandf.co.uk

Trademark notice: Product or corporate names may be trademarks or registered
trademarks and are used only for identification and explanation without intent to
infringe.

ISBN: 9781032306933 (hbk)
ISBN: 9781032306940 (pbk)
ISBN: 9781003306290 (ebk)

DOI: 10.1201/9781003306290

Typeset in Times New Roman
by Apex CoVantage, LLC

Contents

3   NR Layer 2 and Layer 3   32

P. PRASANT, D. SARAVANAN, J. SANGEETHAPRIYA,
AND MORESH MUKHEDKAR

4   4G and 5G NR Core Network Architecture   46

K. GOWRI, V. KAVITHA, ABOLFAZL MEHBODNIYA,
AND SUBRATA CHOWDHURY

5　5G—Further Evolution　　　　　　　　　　　　　62

AVANTHICA SRI M M, CHANDRAKUMAR T, GAUTAM SRIVASTAVA,
AND SUBRATA CHOWDHURY

NALLAKARUPPAN M K, SIVA RAMA KRISHNAN S, RAMYA G, ABDUL
REHMAN JAVED, ISHITA JOHRI, AND SWETA BHATTACHARYA

TRONG-MINH HOANG, THU-TRANG NGO THI, HONG-HUE NGUYEN THI,
DUC-TAN TRAN, AND NAM-HOANG NGUYEN

VITHYA GANESAN, VIRIYALA SRI ANIMA PADMINI,
V. ANJANA DEVI, SUBRATA CHOWDHURY, GAUTAM
SRIVASTAVA, AND KASSIAN T.T. AMESHO

Preface

In an era when mobile communication has become an integral part of our daily lives, the convergence of machine learning and telecommunications stands as a remarkable testament to the power of innovation. This edited volume, *Machine Learning for Mobile Communication*, delves into the dynamic intersection of these two fields, presenting a comprehensive exploration of the latest advancements and future possibilities.

Mobile communication has undergone a rapid evolution, transforming from simple voice communication to a complex ecosystem encompassing data, multimedia, and Internet services. Meanwhile, machine learning has emerged as a transformative force, empowering systems to learn, adapt, and improve from data without explicit programming. The marriage of these disciplines has given rise to groundbreaking applications that enhance spectrum efficiency, improve user experience, and pave the way for next-generation wireless networks.

This book brings together a diverse collection of experts, researchers, and practitioners who have dedicated their efforts to unraveling the potential of machine learning in the realm of mobile communication. Each chapter in this volume offers a unique perspective on how machine learning techniques can be harnessed to tackle challenges and seize opportunities in this rapidly evolving landscape.

From optimizing resource allocation and managing network congestion to predicting user behavior and ensuring data security, the chapters herein cover a wide spectrum of topics that highlight the multifaceted nature of this field. Whether you're a seasoned professional, a researcher pushing the boundaries of knowledge, or a student eager to grasp the fundamentals, this book provides a valuable resource for understanding the intricate interplay between machine learning and mobile communication.

As the editors of this volume, our aim was to curate a selection of chapters that not only showcase the present achievements but also inspire future endeavors. We hope that readers will find inspiration in the collaborative efforts of researchers from around the globe and will come away with a deeper appreciation for the transformative potential of machine learning in shaping the future of mobile communication.

We extend our heartfelt gratitude to all the contributors who have shared their insights and knowledge in this book. Their dedication and expertise have been instrumental in bringing this project to fruition.

In closing, we invite you to embark on a journey through the pages of *Machine Learning for Mobile Communication*. May this book serve as a guiding light, illuminating the path towards a more connected, intelligent, and innovative world.

Dr. Sinh Cong Lam
Dr. Chirani Lal Chowdhary
Dr. Tushar Hrishikesh Jaware
Dr. Subrata Chowdhury

About the Editors

Sinh Cong Lam received a Bachelor of Electronics and Telecommunication (Honours) and Master of Electronic Engineering in 2010 and 2012, respectively, from University of Engineering and Technology, Vietnam National University (UET, VNU). He obtained his Ph.D. degree from the University of Technology, Sydney, Australia. He is currently with the VNU Faculty of Electronics and Telecommunications in the UET. His research interests focus on modeling, performance analysis and optimization for cellular networks, and stochastic geometry modeling for wireless communications.

Chiranji Lal Chowdhary is an associate professor in the School of Information Technology & Engineering at the Vellore Institute of Technology (VIT) in Vellore, India, where he has been since 2010. He received a B.E. (CSE) from MBM Engineering College at Jodhpur in 2001 and a M. Tech. (CSE) from the M.S. Ramaiah Institute of Technology at Bangalore in 2008. He received his Ph.D. in Information Technology and Engineering from VIT in 2017. From 2006 to 2010, he worked at the M.S. Ramaiah Institute of Technology in Bangalore, eventually as a lecturer. His research interests include computer vision and image processing. Much of his work has been on images, mainly image processing, computer vision, pattern recognition, machine learning, biometric systems, deep learning, soft computing, and computational intelligence. He has given a few invited talks on medical image processing. Professor Chowdhary is editor/co-editor of eight books and is the author of over forty articles on computer science. He filed two patents deriving from his research. He was selected as one of Stanford University's Top 2% Scientists Worldwide for 2021 and 2022.

Tushar Hrishikesh Jaware holds a bachelor's degree in electronics and telecommunication engineering from North Maharashtra University, Jalgaon, Maharashtra, India. He further pursued a master's degree in digital electronics and obtained a Ph.D. in medical image processing from Sant Gadge Baba Amravati University, Amravati. He currently serves as the Dean of Research and Development at the R.C. Patel Institute of Technology in Shirpur, Maharashtra. Dr. Jaware possesses over 18 years of invaluable teaching experience.

He is widely recognized as a Ph.D. supervisor in electronics engineering at North Maharashtra University and at Dr. Babasaheb Ambedkar Technological University, Lonere. Furthermore, he has contributed as a member of the Board of Studies in Electronics and Telecommunication Engineering at North Maharashtra University. Demonstrating his innovation prowess, Dr. Jaware holds three international and national patents, along with six copyrighted works. His research findings have been published in sixty-two esteemed research papers featured in renowned international journals and conferences. His expertise in the field has garnered him invitations as a plenary speaker to numerous prestigious events. Dr. Jaware has been bestowed with several accolades, including the Loksatta Tarun Tejankit Award in 2019 and the GIS Young Innovator and Researchers Award (Central India) in 2018, presented by JSR Laboratory, Pune, in collaboration with the Asian Society for Scientific Research. He also received the esteemed Bright Researcher Award from the International Institute of Organized Research in 2017. Additionally, Dr. Jaware has been honored with twelve awards recognizing his outstanding research and academic contributions by various societies. Notably, he has secured research grants from AICTE under the SPICES scheme and under the Unnat Bharat Abhiyaan initiative.

Subrata Chowdhury is working in the Department of the Computer Science of Engineering of Sreenivasa Institute of Technology and Management as an associate professor. He has been working in the IT industry for more than five years in R&D and has handled many projects in the industry with dedication and on time. He has managed projects related to AI, blockchains, and cloud computing for a range of national and internationals clients. He published four books from 2014 to 2019 in the domestic market and internationally with CRC Press and River Publishers; he has also edited two books, one for CRC and one for River. He has also published more than 30 papers and copyrights and patents in his name. He has been a member of organizing and technical programming committees as well as a guest speaker for more than ten conferences and webinars. He has also reviewed and evaluated more than fifty conference papers, journal articles, and book chapters on AI, data science, the Internet of Things, blockchain, and cloud computing for CRC, Springer, Elsevier, Emerald, IGI-Global, and InderScience Publishers. He is the associate editor for *JOE*, *IET*, and other journals. He has taken parts in workshops, webinars, and faculty development programs as the resource person. He has been awarded by international and national science societies for his eminence contributions in the R&D field, and he has received travel grants from and also been a member of the IET, IEEE, ISTE, ACM, and other accreditation bodies.

Contributors

Kassian T.T. Amesho
Institute of Environmental
    Engineering
National Sun Yat-Sen University
Kaohsiung, Taiwan

A Deivasree Anbu
Journalism and Communication
Bule Hora University
Bule Hora, Ethiopia

R. Madonna Arieth
Department of Computer Science
    Engineering (AI)
Sreenivasa Institute of Technology and
    Management Studies
Chittoor, Andra Pradesh, India

Sweta Bhattacharya
School of Information Technology and
    Engineering
VIT
Vellore, Tamil Nadu, India

Subrata Chowdhury
Department of Computer Science and
    Engineering
Sreenivasa Institute of Technology and
    Management Studies
Chittoor, Andra Pradesh, India

Saumya Das
Department of Information Technology
Sikkim Manipal Institute of Technology,
    Sikkim Manipal University
Sikkim, India

Sreejith L Das
Department of Information Technology
Sikkim Manipal Institute of Technology,
    Sikkim Manipal University
Sikkim, India

V Anjana Devi
Department of Computer Science and
    Engineering
Rajalakshmi Institute of Technology
Chennai, Tamil Nadu, India

Ramya G
School of Information Technology and
    Engineering
VIT
Vellore, Tamil Nadu, India

Vithya Ganesan
Department of Computer Science and
    Engineering
Koneru Lakshmaiah Education
    Foundation
Guntur, Andhra Pradesh, India

Shaifali Garg
Amity Business School
Amity University
Madhya Pradesh, India

Ramya Govindaraj
School of Information Technology and
    Engineering
VIT
Vellore, Tamil Nadu, India

K. Gowri
Department of Computer Science with
    Cognitive Systems
Sri Ramakrishna College of Arts and
    Science
Coimbatore, Tamil Nadu, India

Shashi Kant Gupta
Department of Computer Science and
    Engineering
Integral University
Lucknow, UP, India

Trong-Minh Hoang
Posts and Telecommunications
    Institute of Technology
Ho Chi Minh City, Vietnam

Satheesh Kumar Jaganathan
Sri Venkateswara College of
    Engineering and Technology
Chettoor, Tamil Nadu, India

Abdul Rehman Javed
Department of Cyber Security
Air University
Islamabad, Pakistan

Ishita Johri
B. Tech IT School of Information
    Technology and Engineering
Vellore Institute of Technology (VIT)
Vellore, Tamil Nadu, India

Mirdula K
Department of Applied Mathematics
    and Computational Science
Thiagarajar College of
    Engineering
Madurai, Tamil Nadu, India

Monika K
Department of Applied Mathematics
    and Computational Science
Thiagarajar College of Engineering
Madurai, Tamil Nadu, India

Nallakaruppan M K
School of Information Technology and
    Engineering
VIT
Vellore, Tamil Nadu, India

V Kavitha
Department of Computer Science with
    Cognitive Systems
Sri Ramakrishna College of Arts and
    Science
Coimbatore, Tamil Nadu, India

Siva Rama Krishnan S
School of Information Technology and
    Engineering
VIT
Vellore, Tamil Nadu, India

Abolfazl Mehbodniya
Department of Electronics and
    Communication Engineering
Kuwait College of Science and
    Technology
Doha, Kuwait

Moresh Mukhedkar
Department of Electronics and
    Telecommunication Engineering
D Y Patil College of Engineering
Pune, India

Nam-Hoang Nguyen
University of Engineering and
    Technology
Vietnam National University
Hanoi, Vietnam

Viriyala Sri Anima Padmini
Department of Computer Science and
    Engineering
Koneru Lakshmaiah Education
    Foundation
Guntur, Andhra Pradesh, India

Anchal Pathak
Department of Management, College
    of Business & Economics
Bule Hora University
Bule Hora, Ethiopia

Om Prakash
Sri Venkateswara College of
    Engineering and Technology
Chettoor, Tamil Nadu, India

P. Prasant
Khalsa College of Engineering and
    Technology
Amritsar, Punjab, India

J. Sangeethapriya
Department of Information Technology
Saranathan College of Engineering
Trichirappalli, Tamil Nadu, India

D Saravanan
Department of Computer Science and
    Engineering
IFET College of Engineering
Villupuram, Tamil Nadu, India

Mohd Asif Shah
Department of Economics
Kebri Dehar University
Somali, Ethiopia

Gautam Srivastava
Department of Mathematics and
    Computer Science
Brandon University
Brandon, Canada

Sultanuddin SJ
Measi Institute of Information
    Technology
Chennai, Tamil Nadu, India

T Somassoundaram
Department of Information
    Technology
Sikkim Manipal Institute of
    Technology, Sikkim Manipal
    University
Sikkim, India

Nupur Soni
School of Computer Applications
Babu Banarasi Das University
Lucknow, Uttar Pradesh (UP), India

Avanthica Sri M M
Department of Applied
    Mathematics and Computational
    Science
Thiagarajar College of Engineering
Madurai, Tamil Nadu, India

B. Sundaravadivazhagan
Department of Information
    Technology
University of Technology and Applied
    Sciences
Al Mussana, Oman

Chandrakumar T
Department of Applied
    Mathematics and Computational
    Science
Thiagarajar College of Engineering
Madurai, Tamil Nadu, India

Thu-Trang Ngo Thi
Posts and Telecommunications
    Institute of Technology
Ho Chi Minh City, Vietnam

Hong-Hue Nguyen Thi
Posts and Telecommunications
    Institute of Technology
Ho Chi Minh City, Vietnam

Duc-Tan Tran
Department of Electrical and
    Electronic Engineering
Phenikaa University
Hanoi, Vietnam

Chapter 1

Introduction to 5G New Radio

Shaifali Garg, Shashi Kant Gupta,
A. Deivasree Anbu, and Anchal Pathak

1.1 INTRODUCTION

The wireless industry has expanded massively, with a wide range of media-enabled and user-friendly terminals backed by open-source content-creation tools. Network planning and design methods have been developed as a result of this. With the shift from fixed infrastructure to mobile cellular networks, the focus of study has shifted to network design and optimization. Newer generations of networks are guaranteeing users constant service. First, there was mobile voice capability, and the second generation improved on the capacity and coverage issues in mobile voice networks (Vaezi and Zhang (2017). Third-generation systems prioritized higher data rates, as well as multimedia and spread spectrum capability, and the fourth generation supported high-mobility mobile applications for enhanced mobile services. In Figure 1.1, we can see the progression from capacity/quality to heterogeneous environment integration.

With so many new widely used technologies, networks have undergone a paradigm change. Network topologies have been suggested that allow tiny cellular networks to share the unlicensed spectrum used by Wi-Fi without impacting the performance of the Wi-Fi network. In recent years, smartphone usage expanded significantly. In light of the projected seven-fold growth in worldwide mobile data traffic between 2016 and 2021, as shown in Figure 1.2, and the estimated 1.5 mobile devices per capita by that year, it is clear that network connection speed has to be raised (Pratschner et al. (2018). Rapid expansion in mobile communication has been spurred by the convergence of wireless computer networks and personal mobile communication, as well as the development of new mobile technologies and services in a heterogeneous network environment.

Figure 1.3 shows the estimated device market share from 2016 to 2021. Cellular networks underwent significant changes as a result of the influx of mobile devices, applications, and services. As seen in the figure, portable devices are anticipated to expand to 8.3 billion, and machine-to-machine devices are predicted to increase to 3.3 billion (Series 2015). There has been huge growth in mobile and Wi-Fi traffic as a result of the creation and use of new mobile apps by

DOI: 10.1201/9781003306290-1

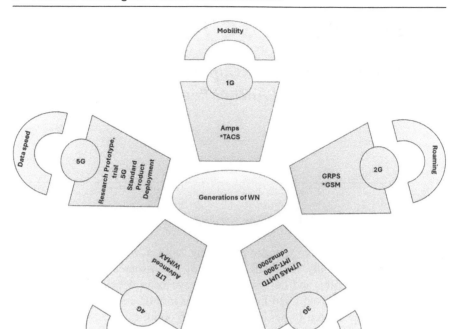

Figure 1.1 The Generations of Wireless Networks.

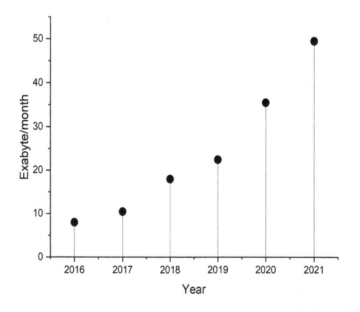

Figure 1.2 Mobile Data Traffic Growth (2016–2021).

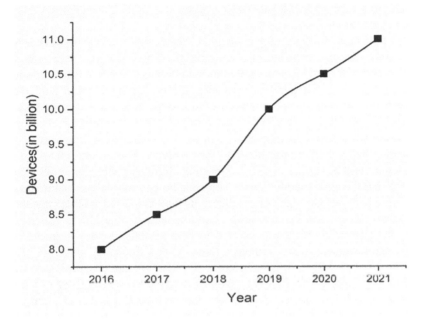

Figure 1.3 Global Growth in Mobile Devices and Connection.

smartphones and tablets. As its use diffuses, 5G will bring together existing cellular networks and other wireless technologies to meet the rising demand for data.

Data quickly supplanted speech as the primary method of communication. Virtual and augmented reality (VR/AR) is an exciting new technology that has been translated into numerous hypermedia formats. Compared with 4G, 5G offers a significant leap forward in mobile communication technology. Smarter 5G network technology will link the whole planet and provide constant access to information, which will change how we live in the future (Bjornson et al. 2019). Already, 5G networks are capable of handling high-speed traffic. It is clear from the technology roadmap that networks, devices, and users have become more diversified. Network components have been blended to provide higher-quality experiences for end users, at the same time conserving resources like energy and radio spectrum. It is possible to support a wider variety of devices with the addition of tiny cells like femto and pico as well as with Wi-Fi. The rest of the research will focus on 5G new radio (NR) in more depth.

1.1.1 Background

A 1000-fold increase in aggregate data throughput over 4G is projected in 5G wireless communication systems, as is an increase in wireless connection dependability. Multiple-input/multiple-output (MIMO) antennas are a key component in making this possible. This is a significant change from previous generations of two- and four-element MIMO antennas employed in mobile terminals (Akhlaghpasand et al. 2017).

One of the most distinguishing elements of the new 5G mobile network is NR, which allows for the inclusion of a wide range of new features, apps, and services. As part of the 3rd Generation Partnership Project (3GPP), NR was designed primarily for use in 5G wireless networks; a more competent and unified 5G air interface and better access methods could result in faster and more responsive mobile broadband experiences that embrace different mobile technologies from various industries. At frequencies up to 52.6 GHz, NR will enhance communication between the transmitter and the receivers. The vision is that 3GPP's modulations will be similar to LTE's orthogonal frequency-division multiplexing (OFDM)-based signal modulations. Ultra-reliable and low-latency communications, massive machine-type communications, and enhanced mobile broadband are some of the criteria for 5G mobile communications that NR is designed to meet. The NR air interface was created to help establish networks in the millimeter and centimeter wave frequency bands with a frequency spectrum below six GHz. Massive MIMO was included in the NR structure. Increasing network capacity and coverage are two key goals of massive MIMO, both of which are necessary to fulfill the soaring demand for data services (Robaei et al. 2020).

1.1.2 Expectations for 5G

The following are a handful of the most anticipated aspects of 5G networks.

1.1.2.1 Outstanding Client Support and Expertise

Customers' requirements for broadband services are growing at the same time as traffic and data usage grows. A more diverse network environment necessitates a more effective administration of network services. In a heterogeneous environment, the emphasis is mainly on user-centric services that provide users freedom of choice in terms of a network.

1.1.2.2 A Steady Connection

There is a pressing demand for rapid service availability and adaptation in the emerging digital age. For this reason, new mobile applications will spring up, pushing today's network to its limits. Machine-to-machine services will be extensively embraced when the network can handle tremendous loads (Sun et al. 2019).

1.1.2.3 To Deal With Unexpected Increases in Network Capacity

It is necessary to optimize storage capacity and power costs to meet rising server workloads and network bandwidth requirements. It is estimated that there are about 1.5 billion web pages, 200,000 Android applications, and 450,000 iPhones in use now, which fuels the need for the Internet of Things (IoT).

1.2 MOVING TOWARD A 5G NETWORK

New characteristics in 5G allow it to merge all previous network generations and fulfill the growing demand for data in a short period. A fully IP-based model is formed by the 5G mobile system user terminal and many network technologies. Frequency licensing and flow rate will no longer be a problem with 5G's innovative technologies. Software-defined radios will be used in 5G, allowing it to accommodate huge channels without the need for streaming. E-health and other new application areas requiring network support will benefit greatly from the next generation of networks (Khalid et al. 2020).

Network quality of service (QoS) has shifted over time, and this will be further enhanced in the next generation of networks. Solutions like software and network intelligence will play a major role in overcoming the problem of network traffic. One of the best ways to improve the performance of next-generation networks is to modify traffic flows based on application QoS requirements. The core concept of 5G anticipates a collaborative effort among all of the network providers. All interconnection expenses will be removed, and the complexity of the system will be decreased. As the number of the network's end-to-end entities is decreased, metrics like latency will be significantly lowered. Converged new technologies such as cognitive radio, cloud computing, and nanotechnology will feature in the new networks. Table 1.1 shows the many generations of wireless cellular networks.

1.2.1 5G Drivers

To justify the large capital expenditures required by operators to launch 5G services, 5G cellular networks must make a major leap in features and performance above LTE. As a reminder, we have a single worldwide mobile phone standard. Following are some inspiring considerations for deploying 5G:

- Improved user data throughput.
- Enhanced system performance.
- Countless connections.
- Reduced end-to-end delay.
- A diverse selection of products and services.
- A wide range of bandwidth options.
- Network adaptability.
- Make the switch to more energy-efficient communication methods.

The International Telecommunication Union objectives for 5G were published in the form of IMT-2020 specifications. IMT advanced criteria were also added

Table 1.1 Generations of Wireless Networks

	1G	2G	3G	4G	5G
Service	Mobile telephony	Packetized variant of a larger capacity	Integrated high-quality audio, video, and data	Electronics equipped with AI	IoT, machine-to-machine communication
Technologies	Analog wireless cell phone technology	Wi-Fi network	Digital broadband network	High-speed connections	"Micro," "femto," and "pico" cells in a multitiered network
Shortfalls	Ineffective, unsafe links	Mobility was a big concern with digital transmissions.	Need for higher network capacity	Rolled out, still establishing	Expected by 2020
Technologies	Analog cell phone tech	Packet data, circuit data	Data packets from a DSL network	High-speed digital packet transmission	Proposed unification of IP and current generating network
Features	Analog signal based **Services:** voice service support, no data service	Digital signals used. Voice service is digitally encoded. **GSM:** voice clarity and integrity **GPRS:** MMS, Internet communication	The network capacity and spectral efficiency have improved. **Services:** video streaming, better Web browsing, support of IPTV	Data and voice over IP converged Greater bandwidth for multimedia services **Services:** mobile TV, high-definition audio and video streaming, IP phone	Access to varied wireless technologies, complete wireless communication **Services:** wearable AI devices, dynamic information access
Data	2 kb/s	6.4kb/s	2 Mbit/s	200–1 Gbps	1 Gb/s and above
Switching	Circuit	Integrated circuit	The remainder of the air interface is packetized and supported by circuits.	Packet switching Message changing	

for comparison. LTE will offer wide area coverage, control and data planes, and connection to a burgeoning evolved packet core (EPC), and 5G NR will enable both nonstandalone and standalone deployments (Elshirkasi et al. 2020). High-speed data will be provided through a dual connection in 5G services. Connectivity to 5G CN will be provided through a standalone deployment of control and data planes. Smart homes, linked drones, eHealth, connected energy, self-driving vehicles, real-time VR/AR games, and more are all possible applications for 5G. LTE has begun using low-latency approaches to help the network transform in anticipation of the vast range of 5G services.

1.2.2   5G Technologies

Today, 5G mobile networks are undergoing extensive research and development initiatives, with commercialization already underway. More than 1000 times more devices per square kilometer are planned to be supported by 5G networks compared with the existing 4G networks, among other criteria. To satisfy these objectives, many techniques have been introduced for 5G networks including massive MIMO, software-defined networking (SDN), multiaccess edge computing (MEC), and radio access network (RAN). Some of these technologies are discussed next.

1.2.3   MIMO Systems

There are more and more people using wireless networks, and they all want the same services at the same time; to respond to this pattern, MIMO systems are becoming more common. MIMO improves data rates and coverage without adding radio frequencies, and a MIMO system's many broadcasts and receiving antennas allow for greater data transfer. MIMO, which uses multiple paths, allows data to be received several times with little latency. The envelope cross-correlation (ECC) between the antennas is used to measure MIMO performance. By lowering ECC, MIMO aims to increase channel capacity. Using MIMO, many independent channels may be created in the original spectrum, and diversity can reduce multipath fading to boost data transmission (Carrera et al. 2020).

1.2.4   Comparison of MIMO in 5G vs. 4G

For 4G mobile networks, 2 × 2 MIMO systems are effectively employed, and for 5G communications, more antenna components are added. Antenna designs for 5G MIMO smartphones have lately been presented in various forms. At frequencies below 6 GHz, 5G MIMO antennas on smartphones must fit into a restricted space and cohabit with antennas from earlier generations. Antennas for 5G MIMO smartphones use a four- or eight-element design, with each element separated from the others by a distance that ensures minimum mutual interaction. This is a challenge for smartphones with little storage capacity. Since 4Gs often take up the space on the short ends of the chassis, the 3500 MHz 5G band antennas on mobile devices are located there. There are various 4G services like LTE and GSM 850/900 that have main antenna sizes of 45 × 5 mm and 5G auxiliary antennas of 19 × 15 mm.

1.2.5 MIMO and 5G

MIMO antennas in mobile terminals provide wider bandwidth and greater data throughput and connection reliability for 5G. MIMO may deliver data speeds over a GB/s without increasing the energy, capacity, or number of retransmission devices needed, all while operating at frequencies below 6 GHz. A rich dispersion environment, on the other hand, is beneficial for MIMO antennas. Research shows that MIMO antennas at cell edges and interior applications are critical for 5G, even though they need high frequency and gain to decrease atmospheric disturbances at millimeter frequencies, a significant challenge.

1.2.6 Massive MIMO and 5G

For artificial intelligence and mobile computing applications, the 5G network is a boon since it requires faster data throughput. Massive MIMO and carrier aggregation with up to 200 MHz BW are employed in 5G NR. This is achieved by providing a large number of autonomous terminals with sufficient narrow beamforming algorithms, such as channel state information (CSI) (Kim et al. 2019).

CSI is a physical layer technology that is used in the wireless sector. Space multiplexing enhances or even replaces the ultra-dense networks capable of handling the large numbers of users in urban and suburban areas. The antenna gains compensate for the propagation loss inherent in the very-high-frequency band of millimeter wave (mmWave) communications. It is possible to place huge antennas at base stations in a scattered or co-located manner. Co-located huge MIMO antennas reduce backhaul costs by consolidating all antennas in a small region. Because the antennas are dispersed over a larger region than in distributed massive MIMO systems, the coverage probability is higher, but the backhaul needs are greater.

1.2.7 Software-Defined Networking

Software-centric networks are being supported by network function virtualization (NFV) and software-defined networking (SDN). Network operators will benefit greatly from these capabilities in terms of both technological and economical aspects, including more manageable network monitoring, improved support for new feature rollouts, and the ability to transfer the network. As a result, a slew of new participants (Internet service providers, cable service providers, etc.) will be allowed to enter the market. First, the EPC was virtualized, and then the physical layer was moved down the protocol software stack toward virtualization.

Network operators can become more flexible in their deployments of diverse use cases by switching to SDN. Network slicing is one of the advantages: Sliced networks are capable of bringing together a wide range of services to meet the needs of specific use cases, and as part of the development of LTE, 3GPP is responding to this desire. SDN and NFV are utilized in network slicing to develop virtual topologies of a network. Based on a common physical infrastructure, these virtual networks may be "tuned" for certain applications, services, or operators. To meet the performance standards, a comprehensive end-to-end logical network is created.

1.2.8 Multiaccess Edge Computing

The 5G air interface enhancements alone won't be enough to meet latency expectations; the network must also be optimized. Core network or data center tasks are moved closer to the network edge, where the data is operated, using multiaccess edge computing (MEC), which has been shown to considerably reduce user end-to-end latency. The change has greatly decreased back-and-forth traffic, which has made it possible to lower the backhaul traffic as well.

MEC makes it possible for cloud computing to be located inside the user's local access network, and fog computing helps with this as well. Antennas in remote radio heads (RRHs) connecting to the RAN are considered the network's edge. At the network edge, there are several reasons to put computer power to use. Reduced server latency (or delay) is the most important consideration for mobile applications (Popovski et al. 2018); with reduced latency, a packet no longer has to wait to be processed when it enters the wireless network. MEC servers closer to the edge will have a reduced impact on applications' performance. While latencies as low as 1 millisecond are required to enable industrial robots and self-driving cars, those as high as 10 milliseconds are required to support virtual reality headsets, and those as high as 100 milliseconds are required for assistive driving systems. Distributing CN and data center functionality to the edge is shown in Figure 1.4.

Additionally, it is possible to reduce backhaul traffic by avoiding retransmitting big packets back to the edge after they are processed in the network. Market drivers for MEC installations include computation and storage capabilities:

- To reduce the overall cost of ownership.
- Increase profitability by using cutting-edge technology.
- Extending virtualization to the access network is logical.
- Boost productivity.

What is the point of having the network's edge at the antenna? We should abandon the network/device black-and-white dichotomy in favor of a more colorful, fuzzy-bordered one. The overall number of wireless devices is predicted to exceed 20 billion by the year 2025, according to several studies. As the world's population continues to grow, so does the number of gadgets. Another point to consider is

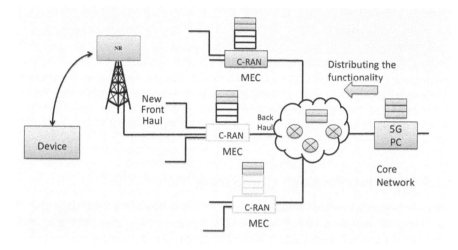

Figure 1.4 Network Diagram Showing Edge Computing Distribution.

that as processing performance on devices becomes more sophisticated and powerful over time, electronics should be seen as the network's edge.

1.2.9   Radio Access Networks in 5G

Distributed RAN (D-RAN; Figure 1.5a) and C-RAN, which stands for centralized, clean, cloud, and collaborative RAN (Figure 1.5b) are the two most common
RAN designs for 5G on today's mobile networks. Each cell site's baseband and
RRHs communicate with one another through the common public radio interface,
and data exchange takes place via the backhaul in a D-RAN architecture. According to China Mobile Research Institute's C-RAN proposal, most RAN functions
are centrally housed in a resource pool of baseband units (BBUs). In the C-RAN,
the base station is divided into BBUs and radio heads, and a high-performance
data center with digital signal processor (DSP) hosts the BBU. The fronthaul
transport network connects the BBU to the RRHs through a high-speed optical
fiber. Transforming a digital signal into an analog signal for transmission is the
job of the RRH.

C-RAN has developed into newer virtual RAN (V-RAN) and open RAN
(O-RAN) technologies. However, V-RAN virtualizes BBUs installed on multiple conventional ×86 hardware platforms at central data centers, whereas RRHs
remain at their separate base stations. V-RAN is a critical component of O-RAN,
as shown in Figure 1.5c. To automate network operations, O-RAN uses artificial
intelligence and machine learning (Chaudhary et al. 2020).

As an alternative, a different RAN split can be used in conjunction with
an alternative protocol that can make use of greater bandwidth technologies
like ethernet-based protocols. A variety of RAN split options (recommended

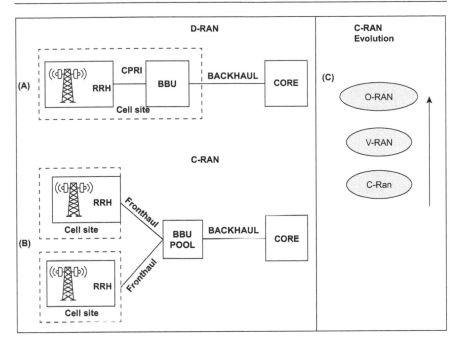

Figure 1.5 (A) D-RAN Architecture, (B) C-RAN Architecture, (C) C-RAN Evolution to V-RAN, and O-RAN.

by the 3GPP) can reduce fronthaul bandwidth consumption and latency. One RAN split option conveys modulated symbols, which is a point in the processing chain before the iFFT procedure on the transmit side converts them to the time domain. More carrier-antenna combinations can be accommodated due to the decreased frequency domain sampling rate. Even though the processing is decentralized, this method allows for more complex scheduling among cells. User data packets, such as packet data convergence protocol packets, are also transported using the RAN split option. To address any security issues, these packets have their headers compressed and correctly ciphered and secured. As a consequence, the data transfer rate is significantly reduced, but the centralized processing capability is lost.

Control and user planes are separated to allow for different evolution rates, decreased latency, and support for novel deployment situations, in addition to dividing RAN operations. An LTE macrocell provides the control plane, while the user plane can be supplied by a small cell 5G. For example, for the 5G cellular system, the time slots are specified to be changeable to manage the vast range of service needs. Spectrum, as seen in 4G, is critical to delivering high data speeds. To accommodate different spectrum deployments, the OFDM access parameters are variable.

1.2.10   Frequency Bands in 5G

With the sub-6 GHz spectrum, a macrocell base can be used for countrywide deployment with a balance between high transmission rate and minimal latency for end users. LTE bands 42, 43, and 46 make up the majority of the available spectrum. Antennas on mobile devices take advantage of the short wavelength of 3500 MHz, which is 86 millimeters in the open air. Sub-6 GHz frequency ranges are very congested, with just a small amount of available capacity. As a result, many mmWave frequency bands have been proposed and authorized in most industrialized nations to satisfy the high-performance data transmission needs of 5G. There is no need for a worldwide license for frequencies between 24 and 29.5 GHz, 37 and 42.5, 47.2 and 48.2 GHz, or 64 and 71 GHz. The development of low-cost, energy-efficient, high-gain broadband antennas that can overcome propagation losses is a problem for the usage of mmWave in 5G. The majority of the mobile device's mmWave antennas are situated on the side of the device (Zhang et al. 2017).

1.3   FREQUENCY BAND ADVANTAGES AND DISADVANTAGES

High-bandwidth, low-latency networks can be achieved using mmWave, but its high frequency and short wavelength make it vulnerable to atmospheric absorption, body absorption, and environmental obstructions like raindrops, snowflakes, and sand; its other drawbacks include its limited range without sight, poor diffraction capability, loss of surface waves, metallic losses, and path loss. Spatial multiplexing can overcome the negative impacts of the mmWave channel by implementing dense networks with many antennas (MIMO) in tiny cells. It's possible to increase the number of devices connected to the network as well as the data rates per user by deploying more small cells in a particular location. Because each mobile operator must set up and run its base station, tiny cells can't serve multiple operators or accommodate a wide range of services. Table 1.2 lists the benefits and drawbacks of 5G's various frequency bands.

1.4   CONCLUSION

Here, we described different generations of network systems to expand the understanding of the present need for current-generation networks in terms of increasing traffic and integrating the expanded range of devices in the new-generation network system. We examined the introduction of 5G NR and explored its progress and expectations. The coming 5G wireless networks feature several critical technological components including massive MIMO, software-defined networking,

Table 1.2 Advantages and Disadvantages of 5G Band Frequency

Band Freq	Benefits	Limitations
Less than 1 GHz	Higher frequencies (1–6 GHz) provide less coverage than lower frequencies (mmWave bands up to 66 GHz).	Compared with other frequencies (1–6 GHz), this one has a lower channel capacity (bands up to 66 GHz).
1 to 6 GHz	The initial wave of 5G deployments is fueled by a well-balanced mix of coverage and channel capacity for 5G services.	In the region of 1 to 6 GHz, there are alternative mobile bands that might be progressively reallocated for usage in 5G.
Greater than 6 GHz	Lower than other frequencies (1–6 GHz) and less reliable coverage. Unlicensed mobile bands will be used to provide the ultra-high Internet speeds anticipated for 5G technology.	In comparison with other frequencies (1–6 GHz), its data rate is higher (mmWave bands up to 66 GHz).

multiaccess edge computing, and radio access networks that we addressed here, and we also discussed the benefits and drawbacks of 5G's various frequency bands.

REFERENCES

Akhlaghpasand, H., Razavizadeh, S.M., Björnson, E. and Do, T.T., 2017. Jamming detection in massive MIMO systems. IEEE Wireless Communications Letters, 7(2), pp. 242–245.

Bjornson, E., Van der Perre, L., Buzzi, S. and Larsson, E.G., 2019. Massive MIMO in sub-6 GHz and mmWave: Physical, practical, and use-case differences. IEEE Wireless Communications, 26(2), pp. 100–108.

Carrera, D.F., Rosales, C.V., Azpilicueta, L. and Galaviz-Aguilar, J.A., 2020. Comparative study of channel estimators for massive MIMO 5G NR systems. IET Communications, 14(7), pp. 1175–1184.

Chaudhary, P., Kumar, A. and Yadav, A., 2020. Pattern diversity MIMO 4G AND 5G wideband circularly polarized antenna with integrated LTE band for mobile handset. Progress In Electromagnetics Research M, 89, pp. 111–120.

Elshirkasi, A.M., Al-Hadi, A.A., Soh, P.J., Mansor, M.F., Khan, R., Chen, X. and Akkaraekthalin, P., 2020. Performance study of a mimo mobile terminal with upto 18 elements operating in the sub-6 ghz 5g band with user hand. IEEE Access, 8, pp. 28164–28177.

Khalid, M., Iffat Naqvi, S., Hussain, N., Rahman, M., Mirjavadi, S.S., Khan, M.J. and Amin, Y., 2020. 4-Port MIMO antenna with defected ground structure for 5G millimeter wave applications. Electronics, 9(1), p. 71.

Kim, J., Sung, M., Cho, S.H., Won, Y.J., Lim, B.C., Pyun, S.Y., Lee, J.K. and Lee, J.H., 2019. MIMO-supporting radio-over-fiber system and its application in mmWave-based indoor 5G mobile network. Journal of Lightwave Technology, 38(1), pp. 101–111.

Popovski, P., Trillingsgaard, K.F., Simeone, O. and Durisi, G., 2018. 5G wireless network slicing for eMBB, URLLC, and mMTC: A communication-theoretic view. Ieee Access, 6, pp. 55765–55779.

Pratschner, S., Tahir, B., Marijanovic, L., Mussbah, M., Kirev, K., Nissel, R., Schwarz, S. and Rupp, M., 2018. Versatile mobile communications simulation: The Vienna 5G link level simulator. EURASIP Journal on Wireless Communications and Networking, 2018(1), pp. 1–17.

Robaei, M. and Akl, R., 2020, January. Examining spatial consistency for millimeter-wave massive MIMO channel estimation in 5G-NR. In 2020 IEEE International Conference on Consumer Electronics (ICCE) (pp. 1–6). IEEE.

Series, M., 2015. IMT Vision—Framework and overall objectives of the future development of IMT for 2020 and beyond. Recommendation ITU, 2083, p. 21.

Sun, L., Li, Y., Zhang, Z. and Feng, Z., 2019. Wideband 5G MIMO antenna with integrated orthogonal-mode dual-antenna pairs for metal-rimmed smartphones. IEEE Transactions on Antennas and Propagation, 68(4), pp. 2494–2503.

Vaezi, M. and Zhang, Y., 2017. Cloud mobile networks (Vol. 5, No. 3, pp. 23–37). Springer.

Zhang, H., Liu, N., Chu, X., Long, K., Aghvami, A.H. and Leung, V.C., 2017. Network slicing based 5G and future mobile networks: Mobility, resource management, and challenges. IEEE Communications Magazine, 55(8), pp. 138–145.

Chapter 2

NR Physical Layer

Mirdula K, Chandrakumar T, Mohd Asif Shah,
and Duc-Tan Tran

2.1 INTRODUCTION

LTE, long-term evolution, considerably decreases network latency, the delay encountered during data transfer. For real-time applications where prompt response is needed, such as online gaming, video conferences, and self-driving car communication, low latency is necessary. For mobile devices, LTE is the norm for rapid-connectivity data transfer via wireless networks. This paper [1] examines the LTE and beamforming approach and its evolution toward 4G, with a particular emphasis on the efficacy of spatial multiplexing in highly dense cell deployments. In comparison with preceding LTE, a fourth-generation technology, the fifth generation is intended to offer considerable increases in data speeds, capacity, and latency. The shift from LTE to 5G offers faster speeds, lower latency, and more capacity and compatibility for a broad spectrum of cutting-edge applications; it marks a tremendous breakthrough.

In addition, beamforming is now utilized considerably more frequently, similar to LTE, and supports a wider variety of frequency bands than LTE can. User equipment (UE) is the hardware that individuals utilize to connect to a wireless network. This paper [2] summarizes that the 5G new radio (NR) standardization release 16 enhances features for various verticals and deployment scenarios, low-latency communication, industrial Internet of Things, and positioning. The frame architectures with 5G give users flexibility in the frequency band to use or in optimizing for high capacity versus low network and UE power consumption.

Wireless communication technologies, including 5G, employ the beamforming approach to enhance user experience, boost network capacity, and improve signal quality. It entails directing radio waves during transmission and reception to produce focused narrow beams that aim at certain users or objects. GHz stands for gigahertz, which is a unit of frequency to measure the frequency of electromagnetic waves, including radio and microwave signals. This paper [3] discusses beamforming and precoding/combining in 5G NR cellular systems, especially in sub-6 GHz and millimeter wave (mmWave) frequency bands. It highlights advancements in multiantenna solutions and discusses redesigning reference signals and channel state information acquisition procedures.

DOI: 10.1201/9781003306290-2

Massive multiple-input/multiple-output (MIMO) and beamforming complement each other's strengths. While massive MIMO offers spatial multiplexing, cutting-edge signal processing, and enhanced channel estimates, beamforming helps to reduce interference, concentrate energy, and facilitate user-specific transmission. FR2 band design involves allocating frequency ranges between 24.3 GHz and 52.62 GHz for high-frequency wireless communication applications. This paper [4] examines hybrid beamforming performance in user massiveMIMO systems, analyzing user count, antenna configuration, and modulation type and offering insights for optimizing FR2 band design.

The physical layer's design has changed significantly in the 5G wireless communication technology compared with its forerunners. This paper [5] analyzes channel estimators in 5G NR systems, focusing on least squares, mean squared error, and Kalman filter estimators for maximum ratio combining and regularized zero forcing receivers. It evaluates key metrics, noise, and interference power and identifies optimal combinations for optimal performance. Channel coding is essential to 5G NR for providing reliable and effective data transfer through wireless communication channels.

This paper [6] also discusses polar codes' widespread adoption in academia and industry, their selection as a channel coding scheme in 5G standardization, and their challenges related to designing code families for rate flexibility and low decoding latency. This paper [7] emphasizes the significance of forward error correction, which is crucial for improving wireless communication systems' reliability and efficiency; channel coding focuses on flexibility, high reliability, low cost, and efficiency. The whole system performance of the 5G NR standard is enhanced by the usage of channel-coding algorithms.

Much attention has been directed toward the 5G wireless communication channel-coding method known as polar coding. It was introduced by Erdal Arikan and gained interest because of its capacity-achieving qualities. In comparison with other channel-coding schemes, polar coding offers benefits such as minimal complexity, large coding gain, and proven binary discrete memoryless channel optimality. A key component of 5G networks, mmWave technology enables substantially larger capacity and quicker transmission rates than ever before. Although the technology has several restrictions, namely in terms of surface area and resistance to occlusion, it does function. Carriers and equipment manufacturers like Samsung and Qualcomm have already introduced mmWave technology to customers and demonstrated some of its advantages.

However, mmWave isn't the only part of the spectrum that will be used to create the next-generation networks, despite the fact that carriers love to brag about their fancy new technologies. The transition from LTE to 5G represents a significant advancement in wireless communication, offering faster speeds, lower latency, increased capacity. Low-density parity-check (LDPC) codes are a different strong channel code that are often employed in contemporary communication systems. The utilization of LDPC coding [8], incorporating diverse lifting sizes and two distinct base graph types, adds complexity to the code block segmentation process at the physical layer. In comparison with 4G systems, 5G channel codes call for increased throughput, latency, and dependability. The 5G codes should work similarly to a 4G system by supporting hybrid automated repeat requests (HARQ) for user data as well as dynamic code rates and durations for operational as well as user information.

2.2   THE 5G NR STRUCTURE

The time–frequency grid that houses physical channels and resources is referred to as the physical structure of 5G NR. The cornerstone of the 5G NR structure is the subframe. It is further split into slots and has a set length of 1 millisecond (ms). Depending on the duplexing mode, each subframe can be set up to transport downlink (DL), uplink (UL), or both DL and UL broadcasts. The lowest unit of resource allocation in the 5G NR structure is the slot, which is a period of time inside a subframe. The subcarrier spacing, which is either 15 or 30 kHz, determines how long a slot lasts. The number of symbols that may fit in a slot depends on the particular arrangement.

In this paper, we discuss 5G's enhanced security features, including edge network preservation, network slicing, visualization, and mobile edge computing [9]. We highlight Verizon's use of 5G security tools and privacy protection in eHealth industries. Higher speed is the key benefit of 5G over 4G LTE, mainly because more spectrum will be accessible for 5G, and it employs more sophisticated radio technology. In contemporary communication systems like 4G LTE, Wi-Fi, and 5G NR, orthogonal frequency-division multiplexing (OFDM) is a digital modulation and multiplexing technology that is frequently utilized.

For effective data transmission across wireless channels, especially in settings with frequency-selective fading and interference, OFDM offers a number of benefits. It is used in digital modulation to encode digital (binary) data on multiple carrier frequencies, a common practice in telecommunications. The fundamental unit representing time in the 5G NR structure is the radio frame. To accommodate different frequency bands and channel bandwidths, the total number of subcarriers and their spacing may be changed.

2.2.1   5G NR Numerology

For various frequency ranges, the subcarrier spacing and symbol length are specified by a set of parameters called 5G NR numerology, which transmits data by breaking up the accessible spectrum into reduced subcarriers. Exceptional spectral effectiveness along with minimal latency are achieved by optimizing the subcarrier spacing and symbol duration. The data to be transferred is subsequently modulated onto the subcarriers and sent through the air interface, then extracted from the receiving device after the subcarriers are demodulated.

Massive data throughput, minimal latency, and device compatibility are all possible with 5G NR numerology because of the use of many antennas and sophisticated signal processing algorithms. It aids in maximizing network performance, establishes the optimum amount of data that may be transferred via the air interface, and determines the subcarrier spacing and symbol period for various frequency ranges. The symbol duration is the amount of time it takes for each symbol to be sent, while the subcarrier spacing is the space between adjacent subcarriers. Different frequency ranges may be optimized for different purposes, such as rapid transfer of data or low-latency communication, by changing these characteristics.

By allowing several devices to use one frequency band, 5G NR numerology additionally minimizes interference and improves spectral efficiency. In OFDM, the avail frequency spectra are split into a significant number of tightly spaced orthogonal subcarriers. Multiple data streams can be communicated simultaneously since each subcarrier conveys a percentage of the total amount of transmitted data. Because of their close proximity and selection of orthogonal frequencies, the subcarriers do not interfere with one another.

2.2.1.1   Subcarrier Spacing

In an OFDM system, the term numerology refers to a collection of parameters, especially the subcarrier spacing, symbol duration, and cyclic prefix length, that form the physical layer structure. The most notable distinction between NR numerology and LTE numerology is that NR supports a variety of subcarrier spacing variants.

2.2.1.2   Slot Length

The length of the slots varies according to numerology. Slot length generally tends to decrease as subcarrier spacing increases. The characteristics of OFDM are what cause this tendency.

2.2.1.3   Advantages

- 5G NR makes use of cutting-edge methodologies like OFDM and channel coding algorithms to optimize spectrum usage and boost data transmission capacity.
- It decreases communication lag and offers low-latency connections for quick response times.
- The technology is made to work with a variety of gadgets and satisfy various communication needs.

2.2.1.4   Disadvantages

- Requires higher processing power and is more complex.
- Range of 5G NR networks may be constrained by lower coverage and penetration caused by the usage of higher frequencies.

2.2.1.5 Applications

- Internet of Things and other applications that need for numerous sensors and devices to be linked synchronously are a perfect match for 5G NR numerology's abilities to accommodate an enormous variety of devices.
- Applications for smart cities including traffic control, security, and energy conservation are made possible by 5G NR numerology.
- Superior-quality video transmission and multiplayer games are possible with 5G NR numerology.

2.2.2 Advantages of the 5G NR Structure

- The adaptable structure enables effective utilization of the accessible spectrum, including high-speed data transfer, minimal latency, and device support.
- The use of diverse numerologies enables speed optimization for various use cases.
- A seamless transition to 5G is made possible by the frame structure's backward compatibility with earlier cellular network generations.

2.2.3 Disadvantages of the 5G NR Structure

- The dynamic structure might be challenging to handle and deploy, demanding complicated software and algorithms.
- The employment of several numerologies may cause network and hardware compatibility problems.
- The 5G NR higher-frequency channels may have patchy coverage and be vulnerable to interference from structures and other obstructions.

2.3 5G NR TECHNIQUES

2.3.1 Beamforming

Cellular base stations utilize beamforming, a traffic-signaling system, to determine the most effective data distribution path to a specific user. A crucial technique that improves the effectiveness and performance of wireless communication networks is beamforming in 5G NR. This paper [10] discusses beam management in 5G standardization, addressing challenges in mmWave communications and utilizing beam pair links for DL and UL transmission/reception. Standardized 5G beam management entails adjusting radio wave transmission and reception in order to concentrate signal energy into narrow beams that go in predetermined directions. Beamforming boosts signal quality, coverage, and network capacity by directing beams at intended users or particular locations. Through beamforming, packet movements and arrival times, multiple users and antennas on a large MIMO array can efficiently share a significantly higher volume of data simultaneously.

Cellular signals are easily obstructed by objects and have a tendency to deteriorate over long distances. Beamforming is a widely used technique to address the challenges associated with modern wireless communication systems. Instead of broadcasting in multiple directions simultaneously, beamforming focuses the signal into a narrow beam directed toward the intended user. This approach increases the probability of the signal reaching its destination without degradation and reduces interference for other users. It ensures better signal quality, improved coverage, and increased capacity by optimizing the transmission towards the intended user.

This unique approach to signal transmission has become an essential component of modern wireless technologies, enabling more efficient and reliable communication. Engineers plan to create the wireless network that next smartphone users, VR (virtual reality) gamers, and driverless cars will depend on daily using beamforming and other 5G technology. Researchers and businesses have already raised the bar for 5G by promising users ultra-low latency and record-breaking data rates. Ultrafast 5G service could be available to customers over the next five years if the remaining issues can be resolved and all these technologies can communicate with one another. Radio frequency (RF) energy is concentrated into a certain direction using beamforming, allowing the RF beam to travel further in that direction. The beam might not be able to reach the UE if the direction is obstructed, so multiple beams from as many different directions as possible could guarantee that the UE has continuous, flawless coverage. This paper [11] gives a brief explanation about beam sweep and measurement.

2.3.1.1 Operation Processes

2.3.1.1.1 Beam Sweeping

Beam sweeping is a method that sends beams in a burst in each of the predetermined directions at regular intervals. Different beams are sent by the gNodeB (gNB) in various spatial directions, and the UE searches and listens for beam transmissions coming from the gNB. The UE establishes channel quality based on the executed beam sweeps and then communicates with the gNB channel quality data. Additionally, the UE and gNB communicate with other data (such as timing data, configuration data, or the data from analog or digital beamforming). The gNB determines the configuration based on the information it receives, such as configuration information for mmWave network access, information for changing beam sweeping periodicity, and information on overlapping coverage for anticipating a handoff to another gNB.

2.3.1.1.2 Beam Determination

The UE then chooses what it considers to be the best beam (or group of beams) to create a directed (and completely beamformed) connection.

2.3.1.1.3 Beam Reporting

After choosing the optimal beam from the gNB, the UE then uses a particular physical random-access channel resource assigned to each DL beam to inform the choice.

2.3.1.1.4 Beam Steering

Beam steering is the process of changing an electromagnetic wave or particle beam's direction. It is frequently employed in industries including optics, radar systems, wireless communication, and telecommunications. Multiple separate, individually controllable antennas or antenna pieces make up antenna arrays or phased arrays. The overall emission pattern of the array can be changed as in Figure 2.1, effectively guiding the beam in the desired direction. Optical beam steering includes directing the propagation of light using tools like lenses, mirrors, or gratings. The direction or focus of the light beam can be changed by adjusting the size, curvature, or placement of various optical elements.

2.3.1.2 Advantages of Beamforming

- Improved signal strength: Beamforming concentrates transmitted energy in a specific direction, resulting in stronger signals. This enhances signal quality, extends the signal range, and improves overall performance in wireless communication.
- Enhanced coverage: By directing energy toward the intended receiver, beamforming expands coverage. It enables better signal penetration through obstacles, reduces signal degradation, and ensures reliable communication over a wider area.

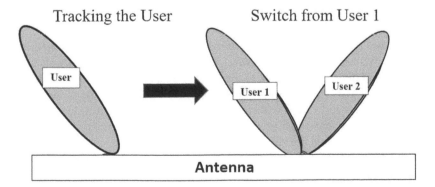

Figure 2.1 Beam Steering.

- Superior signal quality: Beamforming minimizes interference and noise by focusing transmitted energy toward the desired receiver.
- Increased capacity: Beamforming techniques, like multiuser beamforming, enable serving multiple users simultaneously. By creating independent beams for different users or devices, beamforming optimizes spectrum sharing and enhances system capacity.
- Energy efficiency: Beamforming allows precise targeting of transmitted energy, minimizing wasted energy in unwanted directions. This improves energy efficiency, extending the battery life of wireless devices.
- Flexibility and adaptability: Beamforming can dynamically adjust to changing conditions, such as receiver movement or obstacles. This adaptability ensures optimal signal quality, even in challenging environments.

2.3.1.3   Disadvantages of Beamforming

- Cost, hardware, and energy issues from the processing demands.
- Energy efficiency issues exist because omnidirectional broadcasting and communication utilize much more electricity.
- Because beamforming requires heavy computation and power, end-use devices require bigger battery capacity.
- Reduced network coverage since users must be positioned along the beam's direction signals to move in a line of sight.

2.3.2   Massive MIMO

Massive MIMO is a cutting-edge wireless communication method that dramatically improves system performance by making use of antennas on both the server and client sides. It is seen as a crucial enabler for wireless networks of the future, including 5G and beyond. In massive MIMO systems, the base is equipped with hundreds of antennas, which makes possible the simultaneous transmission and receipt of many data streams to many users within the same resource. Massive MIMO's improvement in spectra efficiency and capacity is its key benefit.

With the use of spatial multiplexing, numerous data streams are simultaneously delivered across various spatial pathways. Additionally, massive MIMO enhances link quality and dependability. Due to the huge antenna array's ability to perform highly concentrated and directional beamforming, signal strength is increased while interference and fading effects are reduced. Massive MIMO boosts the signal-to-noise ratio and offers greater coverage and service quality by focusing the transmitted energy on the target consumers.

Massive MIMO also has advantages for energy efficiency. When highly focused beams can be produced, wasted energy can be reduced, and energy resources can be used more effectively since the transmitted power can be concentrated where it is needed. Massive MIMO also has the benefit of supporting many concurrent users. Even in crowded and congested network conditions, the base station can serve multiple users with high data rates and enhanced performance. Numerous applications, such as cellular networks, are being investigated and deployed using massive MIMO technology.

2.3.2.1 Spatial Multiplexing

Increasing the storage and data rate of wireless communication by the simultaneous transmission of numerous independent data streams across various spatial pathways is known as spatial multiplexing, which uses multiple antennae at the server and client. In spatial multiplexing, the transmitter assigns numerous data streams to different antennas, encrypts them, and transmits them all at once. The receiver separates and decodes the signals in order to recover the original data streams. Spatial multiplexing offers faster data speeds and improved spectral efficiency by delivering many streams concurrently.

2.3.2.2 Advantages of Massive MIMO

- Massive MIMO boosts a wireless communication network's capacity. As an example, the sub-6 5G standard permits the deployment of higher frequencies.
- The synchronization of multiple antennas with less processing and minimal power consumption offers exceptional spectral efficiency.
- Additionally, it enhances 4G networks' deep coverage.
- This technology relies on cutting-edge antenna technologies on both access points and end-user devices.
- A massive MIMO system's numerous antennas enable 3D beamforming.
- MIMO improves electromagnetic radiation propagation.

2.3.2.3 Disadvantages of Massive MIMO

- Massive MIMO systems are considerably more comprehensive than conventional base station units and antenna technologies, making them much more expensive. Additionally, the assembly and installation of numerous antenna systems for cellular networks are more difficult and time-consuming due to their more complicated design.
- Frequency division duplex (FDD) produces feedback overhead. This happens when a transmitter receives feedback signals from a receiver. The overhead increases further as the antenna components are increased. Time division duplex (TDD) is hence a better choice for massive MIMO implementation.
- Advanced components that can give their desired degree of performance despite being smaller than their bigger counterparts are required for a complete massive MIMO system.

- The process of developing, creating, and deploying a whole system is costly and complicated enough without the addition of sophisticated processing algorithms.

2.3.3 Channel Coding

Advanced components that can give their desired degree of performance despite being smaller than their bigger counterparts are required for a complete massive MIMO system. Adding redundancy to the sent data through channel coding (Figure 2.2) makes it more tolerant to failures in the communication channel. This paper [12] examines 5G NR systems using cyclic redundancy check-aided polar coding, incorporating channel code rate matching for error correction. It models quadrature amplitude modulation in additive white Gaussian noise and provides recommendations for optimal performance.

In wireless communication systems, channel coding is crucial because channel impairments like noise and fading can affect the sent signal. Without channel coding, a slight mistake might render the received signal useless. By including redundancy into the data, channel coding seeks to lower the error rate of the signal that is received. The receiver can fix mistakes and restore the original data due to this redundancy.

In the realm of 5G NR, both LDPC codes and polar codes [13] play essential roles. LDPC codes are particularly attractive for high-speed data information transfer, thanks to their implementation advantages, especially at gigabit-per-second (Gbps) data speeds. As shown in Figure 2.2, unlike the LDPC codes utilized in previous wireless technologies, the LDPC codes chosen for 5G NR exhibit a rate-compatible structure, allowing for HARQ operation and transmission at various coding rates.

For control signaling in the physical layer, where the information blocks are smaller than with data transmission, 5G NR employs polar codes. To achieve

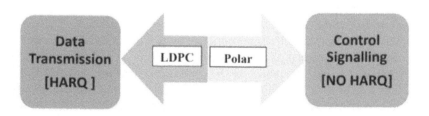

Figure 2.2 Channel Coding.

optimal performance with shorter block lengths, a concatenation approach is adopted. Reed–Muller codes are utilized for the smallest control payloads. By utilizing polar codes for control signaling and LDPC codes for data passing, 5G NR ensures efficient and reliable communication in mobile broadband services. The rate-compatible structure of LDPC codes allows for flexible coding rates, while the concatenation and decoding techniques employed with polar codes optimize their performance for control signaling applications.

2.3.3.1 Techniques

2.3.3.1.1   LDPC Codes

Error-correcting codes known as LDPC codes are frequently employed in communication systems to reduce the effects of channel noise and guarantee accurate data transfer. This paper [14] presents an efficient encoding method and high-throughput low-complexity encoder architecture for quasi-cyclic LDPC (QC-LDPC) codes in 5G NR. QC-LDPC reduces memory storage requirements and hardware complexity, achieving throughputs ranging from 22.1 to 202.4 Gbps. A sparse parity-check matrix with the vast majority of its entries being zeros characterizes LDPC codes. Algorithms for encoding and decoding data are made efficient by this property of sparsity.

Investigators presented a field-programmable gate array-based LDPC accelerator for 5G NR deployments, aiming for the real-time execution of demanding computational blocks, particularly LDPC decoding [15]. The accelerator achieved high decoding throughput and faster execution times while addressing challenges due to lower computational complexity. In LDPC codes, extra bits are added to the original data using a parity-check matrix, enabling the receiver to identify and fix transmission mistakes.

2.3.3.2 Polar Codes

Polar coding is utilized in 5G NR for both the control and data channels. In 5G NR, signaling between the base and the client equipment takes place across the control channels. User data is sent over the data channels. To safeguard the transmitted data from channel imperfections, polar coding is utilized on both channels.

One particular polar code used by 5G-NR's control channels is cyclic redundancy check (CRC)-aided polar (CA-polar) code. Using this code shields the control signaling from mistakes, guaranteeing reliable communication between the base and client equipment. The CRC code finds the errors, and the polar codes fix them. CA-polar codes offer a coding increase of 0.5 decibel (dB), which raises the control signaling dependability.

Different polar codes are used for the data channels in 5G NR than in the control channels. Rate-matching pattern (RMP) polar coding is the type of polarization that is utilized for data channels; the RMP algorithm shields user data from channel degradations. RMP polar codes are created by rate matching a polar code with its user data. The rate-matching procedure includes choosing a portion of polar code bits that corresponds to the amount of the user data. The user data and the chosen bits are then XORed to create the encoded data.

RMP polar codes offer a coding increase of 1.5 dB, increasing the user data transmission's dependability. They are made to handle a variety of coding speeds, enabling effective utilization of the available bandwidth. A greater coding rate produces more dependability but a lower data transmission rate, and it determines how much redundancy is added to the user data. Faster-order modulation techniques produce faster data transmission speeds but worse dependability since they need more bits to transmit each symbol.

2.3.3.3   Advantages

- Low encoding and decoding complexity.
- Power dissipation.
- Less difficult to implement, due to simplified encoding and decoding techniques.
- Coding gain.
- Excellent hardware productivity.
- High throughput (Huawei attained 27 Gbps during 5G field tests).
- Appropriate for control channels with relatively small payload sizes.

2.3.3.4   Disadvantages

- In comparison with LDPC and turbo coding methods, successive cancellation decoders, which are utilized in polar coding, performs badly.
- It provides more O(N) latency.
- Traditional hardware description language-based LDPC decoders are difficult to construct, take a long time, and struggle to fulfill the real-time demands of communication systems.
- Higher cost than LDPC (at finite N).

2.4   MMWAVE COMMUNICATION

The 5G NR standard, which is the recent wireless communication technology, includes mmWave communication as a crucial part, that is, millimeter-wave frequencies, which are commonly found in the radio spectrum between 30 GHz and 300 GHz. Large quantities of bandwidth are available in this high-frequency region, allowing for the transmission of enormous volumes of data at extremely fast rates. Authors of one paper discussed mmWave frequency band usage in 5G mobile networks, focusing on data rates, directivity, propagation loss, and reliability [16]. They address challenges like anti-blocking, dynamic constraints, and device architecture. Recent research on mmWave propagation link characteristics is also presented.

Millimeter wave communication is essential for delivering 5G NR's greater capacity, higher data rates, and low latency. It enables the operation of multiple concurrent connections and the allocation of broad frequency bands, significantly increasing network capacity. Although mmWave and 5G are sometimes used interchangeably, there are significant distinctions between the two; specifically, mmWave is a technology that 5G networks can employ, not 5G itself. Briefly, high-frequency bands can transport far more data but cover much smaller regions than lower-frequency bands, which in turn cover much longer distances but provide slower transmission rates. Although mmWave is only one aspect of 5G, carriers like it because it offers incredibly high bandwidth and boasts the fastest transmission speeds. Increasing the data capacity accessible over smaller, highly populated regions is the aim of mmWave technology.

It will be a crucial component of 5G in many cities, powering data in arenas, shopping centers, conference centers, and pretty much everywhere else where data congestion may be an issue. Sub-6GHz and low bands below 2GHz are particularly important in outlying towns and villages for guaranteeing reliable service across wide regions. Very high-frequency 5G signals have poor indoor-outdoor transitional performance and don't go very far, but precise line-of-sight is not necessary to employ mmWave thanks to massive MIMO and beamforming. Even though mmWave signals can't go very far inside of buildings, they can bounce off of them to maintain a strong signal. Researchers designed a compact, planar four-port UWB MIMO antenna for 5G mmWave applications that featured hexagonal-shaped slotted ground planes and microstrip line feeds, providing high inter-element isolation. The antenna has a 30.47% impedance bandwidth and excellent diversity characteristics [17].

2.4.1 Advantages

- Millimeter wave communication has incredibly large bandwidth and focused beam. With a frequency range of 27–310 GHz and a bandwidth of up to 274 GHz, mmWaves have more than 10 times the bandwidth of microwaves and DC combined.
- Strong detection capabilities and strong security.
- All-weather communication, minimal component size, and superior transmission quality.
- Faster data transfer rates, higher capacity, and significantly lower network latency.

2.4.2 Disadvantages

- Millimeter frequency specification-compatible devices consume a lot of power, especially when linked to a 5G network.

- More money must be spent on mmWave 5G network deployment than on networks according to the sub-6 GHz standard because more smaller cells must be placed in key places to achieve maximum coverage.
- Compatibility is still a problem.

2.5   AIR INTERFACE OF 5G

LTE and 5G have many similarities in their air interfaces and radio protocol stacks (see Figure 2.3). Although 5G provides certain improvements, the names of the four protocols—packet data convergence protocol (PDCP), radio link control (RLC), media access control, and the physical layer—remain the same. Service data adaptation protocol (SDAP) is a protocol that 5G has specified. The core and radio network quality of service (QoS) are connected via SDAP, and SDAP maps the QoS flow for the center network and the data radio bearer that is relevant to the radio network. Data traffic services including robust header compression and ciphering similar to that of LTE are supported by PDCP.

For increased security, 5G also offers integrity protection accessible for data flow. In order to send the PDCP layer, RLC entities from protocol data units (PDUs) are sent to two distinct technologies, LTE and 5G, and the PDCP layer allows support for E-UTRA-NR dual connectivity. Because the PDU copies are forwarded to two distinct carrier frequencies, multiple PDUs increase overall dependability. RLC provides three modes: transparent, unacknowledged, and acknowledged, just as LTE. RLC configuration is carried out per logical channel and is independent of the underlying physical layer structure, including transmission time interval start (TTIS) and numerologies.

To enable ultra-reliable low-latency communications, the MAC layer may impose restrictions on the usage of certain numerology, such as broader subcarrier spacing or narrower TTIS. The MAC layer also multiplexes and schedules logical channels like LTE. While LTE provides synchronous HARQ in the UL and asynchronous HARQ in the DL for greater flexibility, 5G additionally allows asynchronous HARQ in the UL and DL. With support for different numerologies and TTIS, the 5G architecture is extremely adaptable. LDPC and polar coding are used in place of convolutional and turbo coding, and 5G supports lower-frequency bands in addition to massive MIMO and mmWave spectrum.

As Figure 2.3 shows, the physical layer provides a flexible structure and manages multiple tasks including power management, link adaption, and cellular search. The medium-access control layer channels transportation across the air interface and has the critical function of managing data multiplexing and demultiplexing at the RLC layer; it carries out a number of other duties as well, including selecting the appropriate transport format, prioritizing UE, providing scheduling information, and correcting errors. One of its main tasks is also establishing a mapping process between logical channels and transportation channels; by allocating suitable resources and using scheduling procedures, the medium-access

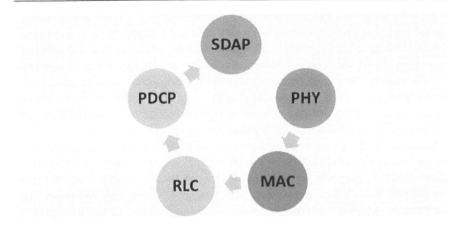

Figure 2.3 Air Interface Protocol Stack.

control layer makes sure that data is transmitted and received efficiently. This enables seamless information transfer between the physical layer and higher layers. The transfer of higher-layer PDUs in line with the designated transmission modes is the main responsibility of the RLC layer, which is also responsible for error correction, sequence numbering, segmentation, and re-segmentation. These processes make sure that data is handled correctly and with integrity while being transmitted.

2.6 CONCLUSION

The physical design of a 5G NR radio structure has been extensively discussed here. The 5G NR structure supports a variety of applications and deployment scenarios since it is versatile and scalable in design. It comprises a radio frame and two subframes, each of which has a single slot. Numerologies supported by the structure are utilized to tailor the network's performance for various scenarios and deploying conditions. In order to fully describe how the structure has evolved, many topics related to the 5G structure were discussed in this chapter.

Beamforming is a crucial basic technique in wide-area networking, where 5G networks are being deployed for smartphones. The subcarrier spacing and symbol period in 5G NR numerology are optimized for various frequency ranges and use OFDM modulation and a configurable frame structure. As a result, there is excellent spectrum efficiency, very little latency, and broad device support. The millimeter wavelength at which 5G frequencies operate makes them more susceptible to interference from obstacles like walls. By enabling a transmitter to aim the broadcast in the desired direction toward a smartphone, tablet, car, or Internet of Things device, beamforming contributes to more dependable communication.

Beamforming will also function with massive MIMO, which directs beams toward customer gadgets in both horizontal and vertical directions at a 5G base station to increase throughput and efficiency. Both the transmitting and the receiving extremities of wireless communication devices can employ signal processing. Prior to modern radar, both receiving and transmitting parabolic antennas had to be mechanically moved and steered to point in the appropriate direction. Signal processing techniques manipulate radio waves and guide them toward a certain location as a focused beam of electromagnetic wave.

Massive MIMO, an antenna technique utilized by 5G NR networks, increases the spectrum productivity, data rates, covered area, and capacity of 5G networks. Massive MIMO supports several user devices at once by using many antenna components (often more than 10 in a single antenna). Massive MIMO arrays that can function at these ranges are the key to 5G throughput and capability, and they will soon become commonplace. The simultaneous development of advanced antenna systems (AAS), which incorporate antenna arrays with the necessary RF transmission hardware and software and signal processing capabilities for beamforming and MIMO, would make it easier to implement these deployments. These AAS will get smaller as mmWave reduces the dimensions of the antennas and electrical components, playing a crucial part in network densification and delivering 5G indoors. In this chapter, polar codes and LDPC codes were also discussed. To be accepted for 5G, however, both codes required to carefully consider concerns related to finite-length code efficiency, complexity, and latency.

REFERENCES

[1] Khan, F., LTE for 4G mobile broadband: Air interface technologies and performance. Cambridge University Press, 2009.

[2] Parkvall, Stefan, et al. "5G NR release 16: Start of the 5G evolution." IEEE Communications Standards Magazine 4, no. 4 (2020): 56–63.

[3] Darsena, D., Gelli, G. and Verde, F., "Beamforming and precoding techniques." arXiv preprint arXiv:2004.05079 (2020).

[4] Dilli, R., "Hybrid beamforming in 5G NR networks using multi user massive MIMO at FR2 frequency bands." Wireless Personal Communications (2022): 1–33.

[5] Carrera, Diego Fernando, et al. "Comparative study of channel estimators for massive MIMO 5G NR systems." IET Communications 14, no. 7 (2020): 1175–1184.

[6] Bioglio, V., Condo, C. and Land, I., "Design of polar codes in 5G new radio." IEEE Communications Surveys & Tutorials 23, no. 1 (2020): 29–40.

[7] Maksimović, M. and Forcan, M., "5G new radio channel coding for messaging in smart grid." Sustainable Energy, Grids and Networks 27 (2021): 100495.

[8] YUAN, Y. and WANG, X., "5G new radio: Physical layer overview." ZTE Communications 15, no. S1 (2020): 3–10.

[9] Mishra, L., Vikash and Varma, S., "Seamless health monitoring using 5G new radio for internet of medical things." Wireless Personal Communications 120, no. 3 (2021): 2259–2289.

[10] Li, Y.N.R., Gao, B., Zhang, X. and Huang, K., "Beam management in millimeter-wave communications for 5G and beyond." IEEE Access 8 (2020): 13282–13293.

[11] Campos, J., "Understanding the 5G NR physical layer." Keysight Technologies release, 2017.

[12] Dilli, R., "Channel code rate matching design in cyclic redundancy check-aided polar coding for 5G new radio uplink communication." Telecommunications and Radio Engineering 80, no. 4 (2021).

[13] Blankenship, Y., Hui, D. and Andersson, M., "Channel coding in NR." 5G and Beyond: Fundamentals and Standards (2021): 303–332.

[14] Nguyen, T.T.B., Nguyen Tan, T. and Lee, H., "Efficient QC-LDPC encoder for 5G new radio." Electronics 8, no. 6 (2019): 668.

[15] Papatheofanous, E.A., Reisis, D. and Nikitopoulos, K., "LDPC hardware acceleration in 5G open radio access network platforms." IEEE Access 9 (2021): 152960–152971

[16] Al-Shammari, B.K., Hburi, I., Idan, H.R. and Khazaal, H.F., An overview of mmWave communications for 5G. In 2021 International Conference on Communication & Information Technology (ICICT) (2021, June, pp. 133–139). IEEE.

[17] Singh, Aditya Kumar, et al. "A compact MIMO antenna for 5G NR frequency bands n257/n258/n261 under millimeter-wave communication." IETE Journal of Research (2022): 1–13.

Chapter 3

NR Layer 2 and Layer 3

P. Prasant, D. Saravanan, J. Sangeethapriya, and Moresh Mukhedkar

3.1   INTRODUCTION

The new 5G networks can be utilized in modern applications such as automobiles, appliances, and even "smart" cities (Figure 3.1). These networks can reach high point rates more than 10 gigabits per second and latencies below one with 10–5 reliability. The advent of 5G will allow for establishing and managing separate

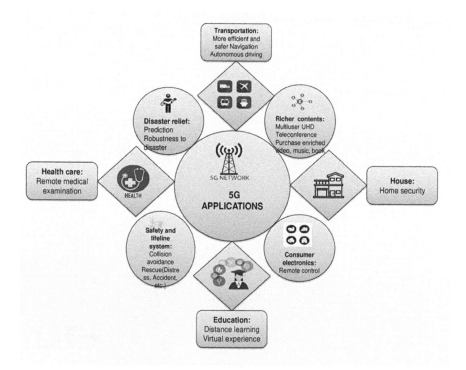

Figure 3.1 Applications of 5G.

DOI: 10.1201/9781003306290-3

logical networks with unique capabilities while benefiting from the economies of scale offered by a public infrastructure (Rupprecht et al. 2019).

3.1.1 Mobile Communication

Rescue workers and police need correct location information in order to respond promptly to emergencies. However, incorporating location information in communications sent via wireless networks presents privacy preservation dangers for users. In terms of driving, for instance, a hostile driver can use location information from monitoring algorithms to track and potentially harm another driver. Or drivers in hit-and-run accidents can spoof their locations so first responders cannot find them and they cannot face legal consequences. Figure 3.2 displays the path of mobile communication (Pérez et al. 2018).

The growth of mobile communication systems (MCSs) has imposed a limit on the radio spectrum's ability to provide ultra-high-throughput services. Optical range, on the other side, may be investigated and will play an increasingly crucial

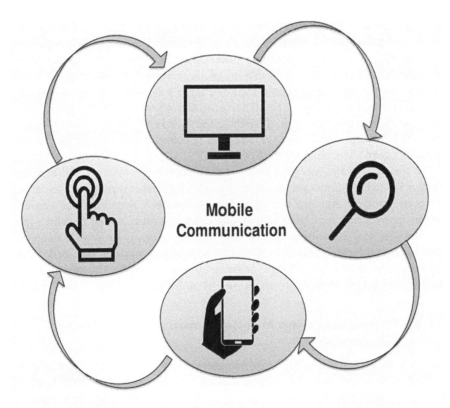

Figure 3.2 Mobile Communication.

part in the future generation of MCSs. An optical wireless signal can achieve near-perfect energy concentration through guided beams, which is vital in the creation of high-throughput wireless communication systems in addition to large accessible bandwidth (Lin et al. 2019).

Energy concentration here refers to directing all possible transmit power in the direction of targeted receivers without wasting energy that would otherwise be propagated to other spatial areas. MCSs have a long way to go in terms of improving their energy concentrations. Smaller cells, for example, can be used to limit the amount of energy sent to smaller areas. One example is the employment of huge multiple-input/multiple-output technology to produce a reduced beam width, allowing energy to be transmitted precisely to appropriate receivers.

3.1.2 Fifth-Generation Mobile Communication

The Internet of Things and micro things concepts require 5G networks to have more system capacity, faster statistics charge in the range of terabits per second, and more users to support them. By 2024, monthly smartphone traffic is anticipated to exceed 136 exabytes, or four times the traffic currently registered in networks. These improvements are based on this data. B5G ultra-fast, dense, and diversified network solutions are becoming increasingly popular as a result (Kim et al. 2017). Multiple orders of magnitude in throughput gain can be achieved through a combination of three essential technologies: great densification of message communications, large amounts of recently obtainable spectrum, and enormous aerial systems that allow throughput gain in the spatial measurement. One way of meeting these extreme needs is to use higher bandwidth beyond the microwave and mm spectra toward frequencies in the terahertz range (Asuquo et al. 2018).

The way access and core networks are developed in 5G systems will demand a major shift. The 5G system, in contrast to previous general-purpose standards, has been carefully designed and optimized to handle a wide range of vertical market services and applications. Network function virtualization, network slicing, smooth local mobility in diverse contexts, and support for a wide range of radio access protocols distinguish 5G systems from their predecessors. In densely populated places as well as in rural areas, satellites are an effective way to complement and extend earthly networks (Yuan et al. 2020).

3.1.3 Fifth-Generation Mobile Communication Using NR Waves

Spectrum flexibility, including high-frequency band operation, interoperability flanked by both lofty and near-ground incidence bands, and lively time division duplex, are all hallmarks of new radio (NR). Supporting these services requires lessening latency and connecting large numbers of devices to the network at the same time. Existing and new services in a wide range of industries will be

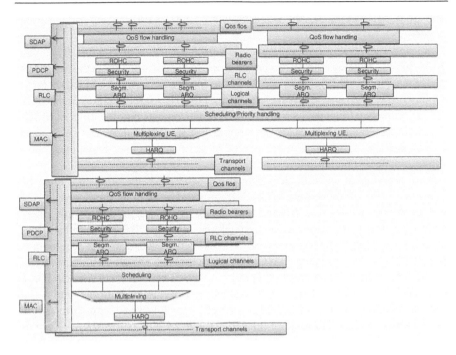

Figure 3.3 NR Radio User Plan and Control Panel.

supported by the 5G mobile communication infrastructure. The convergence of information and communication technologies will be realized through the development of new services. Figure 3.3 shows the NR radio user plan and control panel (Li et al. 2019).

While 5G NR is predicted to give ultra-reduce latency and ultra-wideband services, the 5G mobile communication network is expected to do the same as 4G long-term evolution did a decade ago. Additionally, the 3rd Generation Partnership Project (3GPP) is preparing technical specifications for NR technologies and LTE enhancements in order to enable 5G. There are both benefits and drawbacks to enabling and using 5G NR (Serizawa et al. 2019).

3.2   NR BAND SPECTRUM

The two frequency bands of 5G NR are referred to as FR1 and FR2; FR1 is at 451 to 6000 MHz, and FR2 is at 24.27 to 52.5 GHz. At millimeter wave, FR1 is lower than 6GHz, while FR2 refers to an upper 5G NR band. The sub-6 GHz spectrum is currently used for 4G/3G wireless movable announcement systems, whereas the 5–6 GHz frequencies, which operate at a frequency range of 5150–5825 MHz, are now dedicated to the 5GHz WLAN group. Consequently, 3300–4200 and

3300–5000 MHz of the 5G NR spectrum were created expressly for use in future 5G networks.

FR1 and FR2 core requirements are described in detail in TS 38.101–1 and TS 38.105–2. Notably, mmWave communications will only be possible if NR meets the FR2 standards; the remainder of the requirements will be based on these specifications. FR2 and mmWave are frequently used interchangeably because FR2 encompasses frequencies above 30 GHz. This means that the amount of NR spectrum in FR2 greatly exceeds the amount in FR1. Since these use cases need large capacity and high data rates, as well as an excellent user experience, supporting mmWave operation is a must.

There are several technological challenges with FR2 operation that must be addressed in the establishment of the NR standard as well as the design of the product. A well-thought-out RF and antenna design is required to handle many bands with widely varying carrier frequencies. Both the RF and baseband UE capabilities are put to the test in the carrier aggregation scenario with bandwidths up to 400 MHz and more than 1 GHz. To ensure that this elevated modulation can be maintained, the UE side must efficiently manage a considerable spike in phase noise level for mmWave. This can be reduced by deploying a multielement aerial array on both the system and UE sides, as well as analog beam shaping to ensure that suitable coverage is provided for FR2. Table 3.1 shows the NR FR2 bands (Lopez et al. 2019).

FR2's bands have been precisely established. The UE's performance will be greatly influenced by the form-size factors and specific material characteristics. Integration is more challenging with a smaller device, such as a mobile phone, because it has a smaller screen. Antenna elements can be more easily integrated into larger devices because of the additional space. As a result of this difference, power classes were designed that will hold up to current marketplace demand (Guidotti et al. 2020).

3.2.1 Layers of NR Waves

One of 3GPP's ongoing projects is what they are calling the universal terrestrial radio access network, which aims to provide a smoother transition between a serving cell and a destination cell. Carrier-to-interference ratio and received signal intensity are two common channel parameters used to assess handover. The

Table 3.1 NR FR2 Bands

Sl No	Band Name	F range (MH2)
1	n 257	26500–29500
2	n 258	24250–27500
3	n 260	37000–40000
4	n 261	27500–28350

UE collects and processes downlink measurements of the portion compartment and neighboring cell. A layer 3 (L3) filter is used to remove the effect of fast fading, and layer 2 (L2) assessment issues from the signal. With radio resource control (RRC) signaling, an event reports flipside to the helping bottom position on a intermittent base in the uplink; a handover is started based on the L2 event coverage if the demanded supplies are met (Gohil et al. 2013).

3.2.2 Structure of L2

L2 has a medium access control to multiplex and schedule the data radio bearer (DRB) packet transfers across many transmission channels accessible by L1. Furthermore, L2 includes the amount of processing capabilities that can be customized per DRB, including as segmentation, ARQ retransmissions, compression, and ciphering a new L2 sublayer called service data adaptation protocol (SDAP) was built into the NR specifications to map the data radio bearers (DRBs) and the traffic handled by the 5G. Figure 3.4 shows the L2 structure.

3.2.3 Function of L2

NR L2 is subdivided into the following layers: SDAP, PDCP, RLC, and MAC. SDAP's main function is to map across a quality of service (QoS) flow, which requires a DRB. At the PDCP sublayer for the user, numbering sequences is

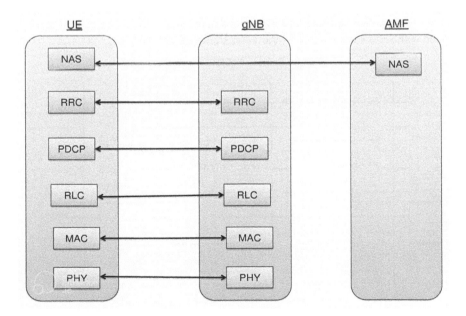

Figure 3.4 Structure of Layer 2.

possible, although compression and decompression of the headers are limited to robust header compression and user data transmission.

The PDCP sublayer for the control plane provides the following functions: numbering in a sequence; integrity protection, ciphering, and deciphering; data transfer from the control plane; detecting duplicates in the PDCP PDU RLC layer (Chettri and Bera 2019). The RLC sublayer's major functions are dependent on the transmission mode and include transferring PDUs from the higher layer independent of the PDCP sequence numbering, correcting ARQ errors, and segmenting and resegmenting service data units (SDUs) that were discarded by the RLC as well as reestablishing the RLC.

The MAC sublayer provides the following functions: MAC SDUs from one or more logical channels are multiplexed and demultiplexed into and out of transport blocks sent to and from the physical layer on transport channels for data-reporting preparation. HARQ is used to repair errors, dynamic scheduling allows for priority management between user devices, and logic channel prioritization manages the priority between logical channels of a single UE and padding (Morgado et al. 2018).

3.2.4 Structure of L3

The radio protocol stack's data transmission services, known as radio bearers, are activated and maintained using the RRC protocol and radio resource management functions like radio bearer control, radio admission control, and connection mobility control. DRBs can be constructed for each user to define the networking service offered by 5GC for each PDU session, as defined above. Most of the blocks that can be inspected for GBR DRBs in the radio admission control can be found within L3 (Series 2017). Figure 3.4 shows its structure.

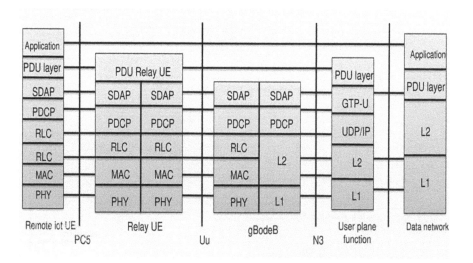

Figure 3.5 Structure of Layer 3.

3.2.5 Function of L3

The RRC sublayer's major services and functions are as follows: broadcasting AS and NAS system information; 5GC or NG-RAN initiates paging; organization, preservation, and discharge of an RRC association between the UE and the NG-RAN; carrier aggregation adding, alteration, and liberation; dual connectivity in NR or between E-UTRA and NR addition, alteration, and release; and key management and safety. Signaling radio bearers and DRBs are reputable, configured, maintained, and released. Handover and context transfer are among the mobility services available, as are UE cell range and reselection, as well as cell assortment, reassortment control, and inter-RAT mobility; QoS management; UE size and coverage control; recognition and recovery of broken radio links; and NAS communication transport from/to UE (Nashiruddin et al. 2021).

3.3 SECURITY PROBLEMS IN L2 AND L3

The time it takes for a packet to be sent from the source to the destination is called user plane latency (in ms). During an unloaded network for a given service, it's how long it takes to get a message from the radio protocol L2 and L3 SDU ingress point of the radio interface to the radio protocol L2 and L3 SDU egress point of the radio interface if the mobile station is active. The point on the cumulative distribution functions of the user throughput that the user sees is the data rate that the user sees. If you want to figure out how many correct bits have been received, you need to look at how many bits have been sent to L3 over a certain amount of time. Handover and context transfer are two of the mobility services that can be used. UE cell assortment and reassortment, and inter-RAT mobility are some of the other services (Morales et al. 2021).

L2 is a broadcast MAC-level network, and L3 is a segmented IP routing network. It is easy for hackers to get into L2 switched environments, such as customer wiring closets in businesses; a typical L2 security risk is that bad actors can get passwords or other important information from people who use the network. Some features that these attacks take advantage of include a switch's ability to learn MAC addresses, the ability to figure out which end-stations are connected through the address resolution protocol (ARP), and the ability of dynamic host configuration protocol (DHCP) servers to assign IP addresses. Open campus networks can't keep the network safe because anyone can get access to any ethernet port and be a hacker. L2 security is important because the OSI architecture was intended to let various communication layers work without knowing each other. With the help of this layer, cybercriminals can gain access to any data they desire (Ganesan et al. 2019).

Here are discussed some common L2 security attacks. "Man in the middle" attacks employ gratuitous ARP-IP host spoofing by DHCP servers and flood MAC addresses (Komol et al. 2019). Additionally, unauthorized users' data can be broadcast over the network through DoS attacks, and unsolicited packets may be stopped or misdirected by network components. In a secure or uncompromised network, MAC addresses are used to build a L2 forwarding table. Devices are identified by their MAC addresses. Switches commonly overwhelm packets containing unidentified MAC addresses, populating the content addressable memory database with source IP address and port information. Each MAC address must be learned in its own memory space. A switch or bridge implements the forwarding, filtering, and learning functions of L2 in this manner. The forwarding table, on the other hand, has a finite number of addresses (Sajjad et al. 2018).

These switches' MAC address-learning capabilities and forwarding behavior are used to flood or overflow a database. As a switch finally learns a MAC address, this attack takes advantage of the hardware constraint. When L2 forwarding restrictions are surpassed, messages overwhelm every port in a virtual LAN, allowing hackers to eavesdrop or sniff Internet connectivity while decreasing network speed. Port security can limit and detect MAC addresses sharing a physical connection, which is a dynamic feature. Packets with source addresses other than those set by the administrator will not be sent when a secure MAC address is provided to a secure port (Wang et al. 2019). Attempting to access a port that has been designated a secure port will result in a security violation, even with the safest, most secure MAC addresses. Using a secured MAC address to reach another secured port is a violation.

3.4   CONTROL MEASURES

L2 switch environments, like one in a company customer wiring closet, are potentially easy targets for hackers. An assault on the network or system users with the objective of obtaining critical data like keys is among the most frequent security concerns inside the L2 area. This is how these attacks work: They use simple protocol-processing abilities, like a switch's ability to learn MAC addresses or end-station MAC address resolution through ARP-RFC 826 or DHCP server IP address assignments (Kim et al. 2009).

DoS attacks prevent legitimate users' information from being transferred over a network by causing network components to stop delivering messages or to incorrectly forward them (Koutras et al. 2020). MAC addresses are typically used to build a L2 forwarding table in a secure or uncompromised network; the MAC address serves as the device's physical location. The standard switch operation is to flood frames with unidentified destination MAC addresses and populate the content addressable memory table with the source address and port of all packets that arrives at the switch (Yang et al. 2020). The number of MAC addresses to be trained is restricted by the switch's recollection capability, and the number of addresses in the forwarding database is limited. Using the switches' natural MAC address learning and forwarding capabilities, flooding or overflowing this table is possible.

Unknown MAC addresses are flooded into the switch in order to take advantage of its intrinsic hardware constraint. This means that packets will be flooded out to all ports in the virtual LAN, making it easier for hackers to spy on network traffic and slowing down the overall network performance (Abusafat et al. 2021). Dynamic port security allows you to restrict and identify MAC addresses of devices that share a physical port. When a secure MAC address is assigned to a secure port, packets with source addresses outside of the authorized group are not forwarded. Ports that have been designated secure are vulnerable to security breaches when an attempt is made to access the port by a device that does not match any of the previously identified secure MAC addresses. Attempting to access another secure port from a station with a secure MAC address on one secure port is considered a violation. In both circumstances, traffic from the offending station is restricted or halted completely. Port security successfully prevents a MAC address flooding attack by limiting the number of allowed MAC addresses on a port (see Table 3.2) (Ioulianou and Vassilakis 2019).

3.4.1   Slicing

Net slicing is a critical component of 5G systems because it enables the deployment of customizable classification behaviors tailored to multiple submission domains on top of a shared network infrastructure. Core and RAN slices are the most common components of a network slice. Table 3.3 shows some of the different protocols (Nebbione et al. 2020).

Table 3.2 Layer Protocols: Issues and Their Solutions

Layers	Protocols	Issues	Solutions
Physical layer	Institute of Electrical and Electronics Engineers 802.15.4	Data transmit attacks	Cipher block chaining–message authentication code algorithm (Koutras et al. 2020)
Physical layer	Bluetooth low energy	Data transmit attacks: header type	Black networks (Yang et al. 2020)
Network layer	Internet protocol version 4 (IPv4/IPv6)	Threats to neighbor discovery protocol	SEcure Neighbor Discovery (SEND) protocol in IPv6 (Abusafat et al. 2021)
Network layer	Recognition of prior learning	Routing and denial-of-service attacks	Intrusion detection systems (Ioulianou et al. 2019)
Application layer	Extensible messaging and presence protocol	Data transit attacks	Lithe solution (Nebbione et al. 2020)
Application layer	Data distribution service	Publish/subscribe model	Real-time communication (Kang et al. 2021)

Table 3.3 Protocols That Analyze Layer Performance

Authors	Protocol	Purpose	Implementation Layer	Technique	Result
(Kim, 2019)	QoS-related Protocols	Minimizing the handover latency	Layers 2 and 3	Edge cell	Proposed pFMIPv6 is more efficient
(Perez et al. 2018)	RAN slicing	"Enhanced mobile broad band services and another slice for providing mission critical services"	Layers 2 and 3	Network slicing	"Guaranteed or non-guaranteed bit rates"
(Mohanty et al. 2016)	Handoff management protocol	"To enhance the handoff performance of mobile IP that is proposed to support mobility management in wireless IP networks"	Cross-layers 2 and 3	"Cross layer handoff management protocol uses mobile's speed and handoff signaling delay information."	"CHMP significantly enhances the performance of both intra and intersystem handoffs."
(Perez et al. 2021)	RAN slice	"An evaluation of the suitability of these parameters for achieving efficient radio resource sharing and isolation between RAN slices is presented when configuring a network for supporting a slice with multiple enhanced mobile broad band services and another slice for providing mission critical services."	Layers 2 and 3	"Non-guaranteed bit rates."	"Efficient radio resource sharing and isolation between RAN slices is presented when configuring a network for supporting a slice with multiple enhanced mobile broad band."

Implementing multiple and distinct RAN behaviors over a surreptitious group of broadcasting capital is demanding mainly because it requires creating and servicing a transfer separation and defense mechanism (Kang et al. 2021).

3.5 CONCLUSION

Today, 5G is a multiservice network that can hold up a broad range of verticals with a wide range of presentation and repair needs. The technology has emerged as one of the most difficult and intriguing areas of wireless research in recent years. NR layer 2 and layer 3 were examined as part of this study, which explores the fundamentals of future mobile communications. The presentation and structure of L2 and L3, security challenges and control methods, and supporting technological drivers in the 5G NR ecosystem are also highlighted. Research gaps and possible future directions are also examined in this paper. Here, we've included some of the most relevant research areas for 5G NR. The effects are felt across a wide range of frequencies. From 1G to 5G, there have been tremendous breakthroughs in the world of mobile communication, and technology for mobile embedded devices will continue to evolve.

REFERENCES

Abusafat, F., Pereira, T. and Santos, H., 2021, April. Roadmap of Security threats between IPv4/IPv6. In 2021 IEEE International IOT, Electronics and Mechatronics Conference (IEMTRONICS) (pp. 1–6). IEEE.

Asuquo, P., Cruickshank, H., Morley, J., Ogah, C.P.A., Lei, A., Hathal, W., Bao, S. and Sun, Z., 2018. Security and privacy in location-based services for vehicular and mobile communications: an overview, challenges, and countermeasures. IEEE Internet of Things Journal, 5(6), pp. 4778–4802.

Chettri, L. and Bera, R., 2019. A comprehensive survey on Internet of things (IoT) toward 5G wireless systems. IEEE Internet Things Journal 7(1), pp. 16–32 (2020).

Ganesan, K., Mallick, P.B., Löhr, J., Karampatsis, D. and Kunz, A., 2019, October. 5G V2X architecture and radio aspects. In 2019 IEEE Conference on Standards for Communications and Networking (CSCN) (pp. 1–6). IEEE.

Gohil, A., Modi, H. and Patel, S.K., 2013, March. 5G technology of mobile communication: A survey. In 2013 international conference on intelligent systems and signal processing (ISSP) (pp. 288–292). IEEE.

Guidotti, A., Cioni, S., Colavolpe, G., Conti, M., Foggi, T., Mengali, A., Montorsi, G., Piemontese, A. and Vanelli-Coralli, A., 2020. Architectures, standardisation, and procedures for 5G Satellite Communications: A survey. Computer Networks, 183, p. 107588.

Ioulianou, P.P. and Vassilakis, V.G., 2019. Denial-of-service attacks and countermeasures in the RPL-based Internet of Things. In Computer Security (pp. 374–390). Springer, Cham.

Kang, Z., An, K., Gokhale, A. and Pazandak, P., 2021. A comprehensive performance evaluation of different Kubernetes CNI plugins for edge-based and containerized publish/subscribe applications. In 2021 IEEE International Conference on Cloud Engineering (IC2E) (pp. 31–42). IEEE.

Kim, H., 2009. An enhancement of FMIPv6 for packet radio networks which supports the QoS provisioning on the MIPv6. International Journal of Digital Content Technology and its Applications, 3(2).

Kim, J., Kim, D. and Choi, S., 2017. 3GPP SA2 architecture and functions for 5G mobile communication system. ICT Express, 3(1), pp. 1–8.

Komol, M.M.R., Podder, A.K., Arafat, A. and Nabeed, T., 2019. Remote sensing global ranged door lock security system via mobile communication. International Journal of Microwave and Wireless Technologies, 5, pp. 25–37.

Koutras, D., Stergiopoulos, G., Dasaklis, T., Kotzanikolaou, P., Glynos, D. and Douligeris, C., 2020. Security in IoMT communications: A survey. Sensors, 20(17), p. 4828.

Li, D., Capone, G. and Malerba, F., 2019. The long march to catch-up: a history-friendly model of China's mobile communications industry. Research Policy, 48(3), pp. 649–664.

Lin, X., Li, J., Baldemair, R., Cheng, J.F.T., Parkvall, S., Larsson, D.C., Koorapaty, H., Frenne, M., Falahati, S., Grovlen, A. and Werner, K., 2019. 5G new radio: Unveiling the essentials of the next generation wireless access technology. IEEE Communications Standards Magazine, 3(3), pp. 30–37.

Lopez, A.V., Chervyakov, A., Chance, G., Verma, S. and Tang, Y., 2019. Opportunities and challenges of mmWave NR. IEEE Wireless Communications, 26(2), pp. 4–6.

Mohanty, S.; Akyildiz, I.F., 2016. A cross-layer (layer 2 + 3) handoff management protocol for next-generation wireless systems. IEEE Transactions on Mobile Computing, 5(10), 1347–1360. doi:10.1109/TMC.2006.142

Morales-Sandoval, M., De-La-Parra-Aguirre, R., Galeana-Zapién, H. and Galaviz-Mosqueda, A., 2021. A three-tier approach for lightweight data security of body area networks in E-health applications. IEEE Access, 9, pp. 146350–146365.

Morgado, A., Huq, K.M.S., Mumtaz, S. and Rodriguez, J., 2018. A survey of 5G technologies: Regulatory, standardization and industrial perspectives. Digital Communications and Networks, 4(2), pp. 87–97.

Nashiruddin, M.I., Rahmawati, P., Nugraha, M.A. and Akhmad, A., 2021, October. Deployment of 5G NR at mmWave frequency for mobile network in Indonesia's market. In 2021 2nd International Conference on ICT for Rural Development (IC-ICTRuDev) (pp. 1–6). IEEE.

Nebbione, G. and Calzarossa, M.C., 2020. security of IoT application layer protocols: Challenges and findings. Future Internet, 12(3), p. 55.

Pérez-Romero, J., Sallent, O., Ferrús, R. and Agustí, R., 2018, April. On the configuration of radio resource management in a sliced RAN. In NOMS 2018–2018 IEEE/IFIP Network Operations and Management Symposium (pp. 1–6). IEEE.

Rupprecht, D., Kohls, K., Holz, T. and Popper, C., 2019. [IEEE 2019 IEEE Symposium on Security and Privacy (SP)—San Francisco, CA, USA (2019.5.19–2019.5.23)] 2019 IEEE Symposium on Security and Privacy (SP)—Breaking LTE on Layer Two, 1121–1136. doi:10.1109/SP.2019.00006

Sajjad, M., Ahmad, A., Malik, A.W., Altamimi, A.B. and Alseadoon, I., 2018. Classification and mapping of adaptive security for mobile computing. IEEE Transactions on Emerging Topics in Computing, 8(3), pp. 814–832.

Series, M., 2017. Minimum requirements related to technical performance for IMT-2020 radio interface (s). Report, pp. 2410–0.

Serizawa, K., Mikami, M., Moto, K. and Yoshino, H., 2019, September. Field trial activities on 5G NR V2V direct communication towards application to truck platooning. In 2019 IEEE 90th Vehicular Technology Conference (VTC2019-Fall) (pp. 1–5). IEEE.

Wang, H., Xu, L., Lin, W., Xiao, P. and Wen, R., 2019. Physical layer security performance of wireless mobile sensor networks in smart city. IEEE Access, 7, pp. 15436–15443.

Yang, J., Poellabauer, C., Mitra, P. and Neubecker, C., 2020. Beyond beaconing: Emerging applications and challenges of BLE. Ad Hoc Networks, 97, p. 102015.

Yuan, Y., Zhao, Y., Zong, B. and Parolari, S., 2020. Potential key technologies for 6G mobile communications. Science China Information Sciences, 63(8), pp. 1–19.

Chapter 4

4G and 5G NR Core Network Architecture

K. Gowri, V. Kavitha, Abolfazl Mehbodniya, and Subrata Chowdhury

4.1 INTRODUCTION

New Radio (NR) stands as the radio access technology (RAT) at the core of the fifth-generation wireless network (5G), as standardized by the 3rd Generation Partnership Project (3GPP). As of the year 2024, 5G has firmly established itself, surpassing initial expectations. NR plays a pivotal role as the universal system for 5G network frequency interfaces, mirroring the evolutionary path of its predecessor, 4G Long-Term Evolution (LTE), built upon orthogonal frequency-division multiplexing. In this transformative era, 5G has redefined mobile communication standards, notably by effectively utilizing spectrum resources, including the liberation of previously underutilized frequencies above 6 GHz for enhanced cellular services [1]. While LTE infrastructure currently supports non-standalone 5G mobile network technologies, the imminent completion of a standalone architecture signifies a significant milestone, symbolizing the independence of 5G from reliance on 4G networks.

The fifth generation of digital networks introduces a new network capable of linking nearly everyone and anything; it was designed with higher multi-Gbps (gigabyte per second) data speeds, low latency, higher dependability, huge communication network capacity, and an improved and more reliable customer experience for maximum users. To achieve these gains, 5G NR uses dual frequencies: Frequency range 1 (FR1) is used for frequency bands inside 410 MHz–7125 MHz, and FR2 is for frequency bands within 24250 MHz–71000 MHz [2].

4.2 THE EVOLUTION OF THE FOURTH- AND FIFTH-GENERATION CORE NETWORKS

4.2.1 Fourth-Generation Communication Systems

The 4G systems are a better version of IEEE's 3G networks, with improved data rates as well as the facility to handle other multipart software services. In 4G systems, innovative LTE wireless telecommunication are employed. They are backward compatible with previous versions, which allows for quicker construction and advances in the networks [3]. LTE telecommunication aligns with the

DOI: 10.1201/9781003306290-4

concurrent broadcast of speech and data, which intensely improves the data rate [4]. Internet Protocol packets may transmit any service, as well as cell phone services. To grow the uplink and downlink capacity, complex modulations and carrier combinations are applied [5].

Main Features of the 4G Systems

- Far faster data rates of up to 1 Gbps.
- Improved mobility and safety.
- Latency reduction for mission-critical applications.
- High-definition video streaming and gaming.
- Voice-over-LTE networks use IP packets for the voice.

Demerits of the 4G Systems

- Costly physical hardware and structure.
- The spectrum frequency band is prohibitive in many countries.
- Higher-end mobile handsets that support 4G technology are pricey.
- Widespread rollout as well as upgrading take time.

4.2.2   Fifth-Generation Communication Systems

No single firm or separate controls 5G; rather, numerous corporations across the mobile network are working to take it toward life cycle [6]. The center of the process is 3GPP, the business organization that produces global standards for 3G UMTS, 4G LTE, and 5G networks. Many critical inventions are being driven by 3GPP throughout all parts of 5G architecture, including the air interfaces to the network service layer. Infrastructure providers, module/device producers, mobile system operators, and service agents are among the additional 3GPP 5G members.

4.2.2.1   Where Is 5G Being Used?

In general, 5G is deployed in three linked facilities: enhanced wireless broadband, mission critical telecommunications, and the massive Internet of Things (IoT). One distinctive feature of the fifth generation is its plan for forward compatibility, the capacity to handle service in the future that is unimagined today [7].

The 5G communication technology will support innovative technologies such as virtual and augmented reality with extra reliable data rates, decreased latency, and reduced cost per bit [8]. With its ultra-reliable, reachable, low-latency networks, 5G will provide new facilities that are able to adjust segments, for example, the remote controllers of key infrastructure, automobiles, and health processes. With the potential of scaling downcast on bandwidth, power, and flexibility, 5G is planned to effortlessly associate a massive number of built-in sensor devices in nearly everything, giving extremely lean and low cost construction options.

4.3   THE ARCHITECTURE OF THE 4G LTE CORE NETWORK

The 4G NR core network, formerly known as the 4G LTE core network, is an essential component of the 4G cell phone communication system. It is created on a packet-switched design and is made up of several network parts that are working together to allow the 4G Internet connection to function. The network enables high-speed data transfer, creative applications, and continuous mobility. It offers the required underlying architecture, capabilities, and connections for mobile users to communicate in an effective and reliable way. The following are some major features of the 4G NR core network:

- Evolved packet core (EPC): The EPC is the main network framework of 4G NR. It is made up of the following important network components that conduct multiple tasks:
  a. The mobility management entity (MME) is in charge of controlling the portability of user devices inside the network. It performs mobile device duties including authentication, tracking, and paging. The MME also handles network resource allocation and allows handovers across base stations (eNodeBs).
  b. Serving gateway (S-GW): The S-GW serves as a bridge between the radio access network (RAN) and the core network. It is in charge of routing user data packets between the eNodeB and the relevant network nodes. The S-GW is also in charge of mobility-related tasks like user plane stabilization for handovers.
  c. Packet data network gateway (P-GW): The P-GW connects the 4G network to external packet-switched networks like the Internet. It handles things like network address allocation, filtering packets, and data service billing. The P-GW guarantees data traffic security and quality of service (QoS).
  d. The policy and charging rules function (PCRF) is in charge of policy enforcement, including QoS regulations and billing consumers. To guarantee effective policy administration and pricing mechanisms, the PCRF communicates with different network parts.
- Interfaces: To ease communication between its many pieces, the 4G NR core network infrastructure employs a number of interfaces:
  a. S1-MME serves as the link that links the MME and the S-GW. It transports mobility administration and session control signaling signals.
  b. S1-U is used to link the S-GW to the eNodeB. It transports user data packets.
  c. S5/S8 connects the S-GW and the P-GW. It speeds up the interchange of data from users' packets and mobility-related data.
  d. Gx is used to link the PCRF to the P-GW. It is utilized for policy management and billing interactions.

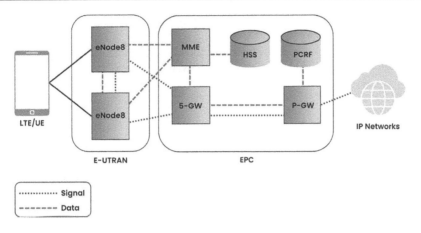

Figure 4.1 4G NR Core Network Architecture.

- Network function virtualization (NFV) and software-defined networking (SDN): NFV enables network services to be constructed using software that runs on standardized hardware, allowing for more flexibility and scalability. SDN divides the management plane and the data plane, providing centralized network control and administration.

The 4G NR core network architecture (Figure 4.1) is intended to deliver high-speed data transmission, innovative services, and effective network mobile management. It provides mobile users with seamless connectivity and communication, establishing the groundwork for a dependable and strong 4G Internet connection [9].

4.4 THE 5G LTE CORE NETWORK

The 5G network employs cutting-edge skills to provide users with ultrafast Internet and multimedia experiences. LTE is the latest network that is going to get upgraded to supercharged 5G networks. Previous distributions of the 5G wireless network work together in both standalone and nonstandalone modes. In the nonstandalone mode, LTE and 5G NR spectra are used concurrently. In nonstandalone mode, command signaling should be connected to the LTE main net [10]. Standalone mode supports a specialized 5G central network with a greater-capacity 5G NR spectrum. The sub-6-GHz FR1 spectrum band has been utilized in the earliest 5G networks.

Main Features of the 5G Technology

- Mobile Internet speeds of up to 10 Gbps.
- Low latency in ms.
- Overall cost savings for data.

- Improves efficiency by utilizing technologies such as tiny cells and beam shaping.
- Forward compatibility network for future advancements.
- Cloud-based architecture provides power effectiveness and easy hardware upgrading.

4.4.1   Network Slicing

Network slicing is a procedure of layering numerous wireless networks over a common network field that gathers linked network and computational resources. The 4G and prior generations of cellular communication networks were unable to handle network slicing [11]. Within the constraints given by the foundational physical networks, every segment of the system can have its own rational architecture, safety standards, and routine characteristics. Changed slices might be allocated to certain tasks, such as ensuring that a single application/service has important access to bandwidth and transfer or separating congestion for particular individuals or device classes.

4.4.2   How Does Network Slicing Work in 5G?

Network slicing (Figure 4.2) is an important aspect of 5G; implementing slices follows the same virtualization concepts as the full provider network architecture. These concepts encompass the networks for radio access and the accompanying transport and mobile primary networks that underpin 5G, as well as the data center resources that go with them. Service providers can use slicing to develop a

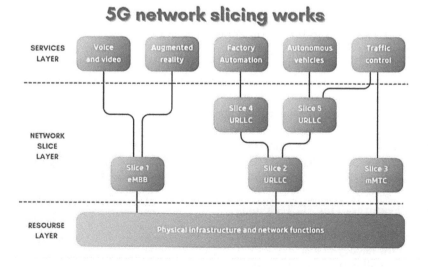

Figure 4.2 5G Network Slicing.

number of network slice topologies that give real assurances to consumers, such as a minimum level of capacity for their links or prioritizing getting packets from specified specific devices or applications [3].

The 5G network's operator can physically separate traffic across wireless connections, slice one network into sections, or integrate the capacity of numerous networks and slice the shared resources. This allows operators to select the features required to meet their intended objectives for spectrum effectiveness, traffic capacity, and connection compactness (the number of strategies that may connect from a given place). The 5G service kinds that employ slicing for asymmetric traffic management consist of the following:

- **Enhanced Mobility Broadband** delivers mobile data access in three ways: to dense groups of clients, to vastly mobile users, and to users dispersed across large regions. It is based on features such as massive arrays of multiple-input/multiple-output (MIMO) antennas, with a mixture of broader spectra than those for ordinary 4G.
- **Major Machine-Type Communication,** or mMTC, services are designed to serve a large number of devices in limited space while generating minimal data (tens of bytes each second) and withstanding significant lag (up to 10 milliseconds on a round trip). Furthermore, the requirements mandate that data transmission and reception consume little energy so that gadgets can have extended battery lifetimes. In a comparable vein, the new 5G radio specification aims for the support of one million devices per square kilometer.
- **URLLC, or ultrareliable low-latency communications,** uses 5G to deliver encrypted communication with speeds of a millisecond (ms) as well as excellent validity with low, or even zero, packet loss. This is accomplished through an array of physical device optimizations on MIMO antenna groups, concurrent frequency bands, packet encoding and processing techniques, and optimized signal handling [12].

4.4.3   Network Slicing Benefits

Slicing, in conjunction with virtual system activities, is essential to "just right" facilities for service workers. Specifically, service providers give only as wide a size for particular services as is needed to fulfill the demands of consumers on a foundation of general radio communication network structure and all-purpose computation and storage resources. Using built-in bulk slicing capabilities and all digital service platforms means that even complicated and unique client environments can be set up quickly. More flexibility to customize affordably delivers the following benefits to service providers:

- They can generate new revenue streams by eliminating the obstacles to experimenting with new service offers. Creating specialized equipment internally eliminates purchasing costs and value depreciation.

- They will have increased flexibility because they will not require dedicated and specialized hardware; 5G will allow more types of services to be delivered concurrently.
- Easier scaling is available because all the physical hardware is generic.
- A higher return on investment is also possible because the capacity to continually test new things allows for the most efficient use of resources [13].

4.4.4   Network Slicing Use Cases

Network slicing facilitates use cases that prioritize performance, capacity, and security. Many of the use cases incorporate new IoT strategies. The following are some examples of network slicing use cases.

4.4.4.1   Performance in Real Time

A corporation that relies on a network using 5G for independent vehicle management could sign for a wireless network slice with delays below five ms and rigorous packet delivery assurances on the lowest speed necessary to provide reactive control. The provider would employ URLLC capability to ensure that the slice is on the lowest-latency infrastructure and pathways connecting automobiles and cloud or edge services, as well as maintaining enough capacity to fulfill the throughput objective.

4.4.4.2   Capacity

Latency may not be an issue for a system that tracks security in a huge auditorium. However, the security firm may choose to purchase assured transmission of the camera's data: The corporation would like 1.5 gigabits per minute of throughput per camera, with no packets missed at any time. To suit the specifications of this slice, the provider may combine several connection and processing device choices.

4.4.4.3   Security

A hospital in a metropolitan region may desire to separate every flow of traffic to and from heart rate monitors for an individual health monitoring system. This would reduce the danger of customer data leakage and the potential of control channel intrusion, necessitating the deployment of a slice containing single-user virtual assets [5].

4.5   5G NEW RADIO

5G NR is a new radiocommunication standard that is faster and more effective, with scalable networks. With 5G NR mobile technology, we will be able to

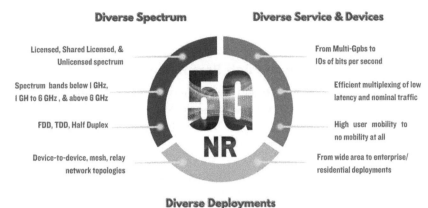

Figure 4.3 5G NR Standard.

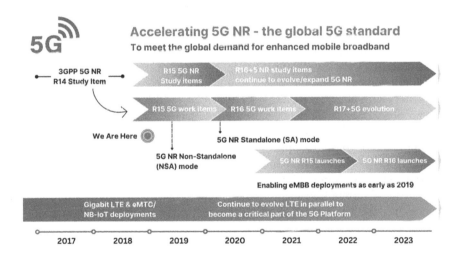

Figure 4.4 The Global Demand for Greater Mobile Broadband.

communicate with many devices in our environments with minimal latency that leads to lightning speed. The third generation of mobile networks allowed for transmitting received data over the mobile network, and the current 4G technology provides greater data rates than earlier generations, but it's limited by capacity, scalability, and the numbers of customers in separate cells [14].

4.5.1   A 5G NR Roadmap

The first 5G NR structure is projected to be introduced in the middle of 2018 or early 2019. But various analyses must be performed, spectrum must be allocated, and structure, including user equipment (UE), must be created within a short time span.

4.5.2 5G NR Tasks with High Performance

Unlicensed spectrum: 5G NR will operate in the millimeter wave (mmWave) band of more than 24 GHz. A few of the 5G NR contenders are bands under 1 GHz, such as 600 MHz and 700 MHz, which are used for long-distance broadcasts. Unlicensed bands ranging from 1GHz to 6GHz are used for increased bandwidth [15].

Millimeter wave: Because of the substantially increased bandwidth, mmWave plays a significant part in the 5G NR system. The following frequencies are top prospects for future 5G networks: from 24.25 GHz to 27.50 GHz, from 27.50GHz to 29.50 GHz, 37 GHz, 39 GHz, and from 57 to 71 GHz.

Massive MIMO: In test circumstances, downlink speeds of up to 952 megabits per second (Mbps) were reached utilizing MIMO technology. To attain maximum speed and efficiency, MIMO employs sophisticated arrays of antennas to multiply transmission speed.

Small Cell: A small cellular network is a collection of low-power transmission base stations that utilize mmWave to increase wireless network size. The overall network burden is distributed across tiny cells, and actual impediments are minimized. Recent network technologies such as beamforming and full-duplex mode may be used effectively for transmission using the tiny network cell idea with mmWave [16].

Dynamic time division duplex (TDD) technology: Provides for effective simultaneous transmission and reception, as well as broader transmission bandwidth. For more efficient spectrum utilization, enhanced modulation systems of current frequency division duplex, TDD, and combination technologies can be applied. A self-contained TDD subframe infrastructure

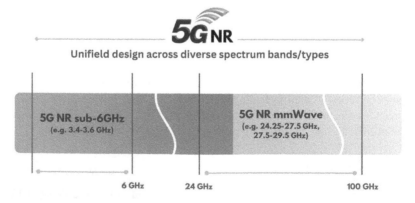

Figure 4.5 5G NR Spectrum Bands.

provides far greater communication capabilities and cohabitation with extra systems supported by the structure [16].

4.5.3   Merits of 5G NR

The 5G NR standard will provide a significantly faster, more scalable, and better organized network capable of supporting the billions of devices with upcoming communication technologies such as the IoT. Furthermore, because of its larger bandwidth of up to 500MHz, unlicensed spectrum provides limitless options. One of the greatest benefits of the 5G NR technology is that it can be improved significantly in the next centuries without disrupting the present networks [17]. Additionally, low-latency architecture provides a more dependable network with more accurate support for mission-important applications, for instance with autonomous driving and linked cars. Low-cost scalable network deployments also offer extensive coverage and high dependability. Because of the lower power usage, these systems are more energy efficient and adaptable.

4.6   5G CENTRAL NETWORK ARCHITECTURE

The new 5G network, as described by the 3GPP network, uses a central, cloud-based, service-oriented design that covers all 5G activities and connections, such as authorization, security, administration of sessions, and traffic flow aggregate from end devices. The 5G base focuses on NFV, utilizing virtualization software services offered by the MEC infrastructures that are critical to 5G structural ideas. For the initial time, UMTS provides a data rate of 384kbps [18].

Earlier cellular network generations' principal purpose was to provide networking customers with fast, dependable mobile data access; 5G has expanded this

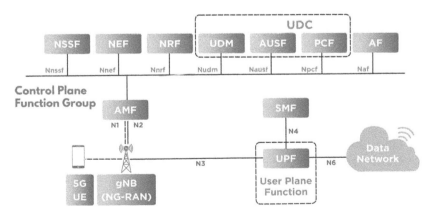

Figure 4.6 5G Core Network Architecture.

choice to provide a diverse variety of radio services to end users across numerous accessing devices and multilayer networks. The technology is a dynamic, cohesive, and adaptable architecture of creative innovations to accommodate a wide range of activities. It employs a smart design, with RANs no longer bound by proximity to improper stations or complicated equipment. With novel interfaces making extra data admission points, 5G surfaces the path for disaggregated, versatile, and dynamic RAN. Next, the main network functions as they are currently defined are described [19].

4.6.1   Server Authentication

It primarily comprises extensible authentication protocol server functions, provides key storage space, and delivers entering data to the sender.

4.6.2   Access and Mobility Management (AMM)

The AMM manages NAS signaling, ciphering, and information safety; registration administration; network and cell management; user authentication; and security. The AMM also serves as the endpoint for RAN CP interfaces and houses the network slice selection function.

4.6.3   Session Management

This function handles session administration (establishing, modifying, and releasing sessions), UE Internet protocol address distribution and management, Dynamic Host Configuration Protocol operations, NAS signaling end associated with session control, DL data statement, and network traffic routing setup for the user plane function for optimal network traffic directing.

4.6.4   User Plane Management

This entails managing packet checking, sending, and transmission and QoS and serves as an exterior PDU assembly point of the connection to the data network and a newscaster point for Intra and inter-RAT mobility.

4.6.5   Network Exposure Management

This function allows for the disclosure of features and instances, the secure transfer of information through outside applications to the 3GPP wireless network, and the transformation of internal or external data. It serves as an application programming interface (API) gateway and permits third-party consumers, such as organizations or partner operators, to monitor, deliver, and maintain application regulations for clients inside the provider's network. As a result, it offers safety and security when network and application functions connect to 5Gen central nodes; it also serves as a proxy and a collection point for APIs, which it otherwise converts into the central network.

4.6.6   Network Repository Management

This function is responsible for discovering network function cases and returning the found instances when it receives a finding demand. It does not exist in 4G. It supports network instance profiles and their supporting services (interfaces, maintenance, administration, etc.).

4.6.7   Policy Control

This function implements the integrated policy framework, delivers policy instructions to CP roles, and accesses contribution data for UDR policy choices for matters including sharing the network and network roaming, as well as network mobility management. It is identical to the existing 4G PCRF.

4.6.8   Unified Data Management

This function requests reader (subscriber) data, outlines and generates authentication and key agreement identifications, and handles verification of users and subscriptions.

4.7   APPLICATION FUNCTIONS

Application functions (AFs) function similarly to a request server that can communicate any of the control plane network functions. AFs can be for various services and be controlled by either the network provider or trustworthy suppliers. For example, the top application provider's AF can impact routing, directing traffic to its external edge servers. The AF may utilize network functions directly for services considered trusted by the operator, but unknown or AFs from outside sources would get the ability to network functions via the network exposure function.

4.7.1   The Data Network

This is the outside information network that allows access to operator services, third-party services, and so on. These new functions of the 5G core architecture inside the control plane such as AMM allow different authorized network operations to connect to their resources. HTTP-based APIs are used to communicate with these nodes, replacing protocols like Diameter. The introduction of stateless functions and API-based interfaces and the movement toward the software world signify significant shifts in the telecom business. The architecture is adaptable and expandable, featuring a multislice UE that connects to numerous services via several slices at the same time, with optimized access and mobility signaling [20]. The APIs provide integration with third-party applications, and the design provides for improved QoS.

4.7.2 5G Core Architecture

The new 5G core, according to specifications by 3GPP, employs cloud-aligned, service-oriented architecture to provide connections, confirmation, safety, session management, and traffic combinations between end devices. The 5G core has an emphasis on NFV, with virtualized software functions provided through the MEC structures [21].

4.6.11.1 Differences From 4G Architecture

The facets of the transition from 4G to 5G include the switching to mmWaves, the huge-scale MIMO, the network splitting, and basically every other architectural change to the multipart 5G eco-systems. The 4G EPC differs greatly from the 5G core, with the 5G core utilizing virtualized as well as software architecture at new levels. The user plane divides packets with the entrance control and user-level functions, and the AMF separates visit organization duties from relationship and transportation administration responsibilities, additional features that separate the core of 5G from its 4G predecessor.

4.6.11.2 5G Architecture Options

Addressing the divide between the 4G and 5G speeds necessitates little stages and a well-planned willing plan. The progressive change from the nonstandalone mode to the independent mode in 5G architectural alternatives is symbolic of this transformation. The 5G nonstandalone specification that was finalized in late 2017 relies on standing LTE RAN and central networks as a newscaster, together with a 5G element provider. Eliminating the dependence on 4G construction, the non-standalone option expands throughput by utilizing millimeter wave bands [22].

4.6.11.3 Topographical Architecture Implementation

Because of the physical components required for solo 5G deployment, a global stage to enable 5G connectivity for multiple geographical locations is required. Strong technological areas such as Asia, North America, and Europe are rapidly scaling up deployment, while other countries throughout the world trail carefully behind. Approximately 200 live 5G networks are now operational throughout the world, with wireless 5G connections expected to hit two billion devices by 2025 [23]. The closeness of neighboring nations, as well as the multiplicity of transporters, have made the deployment in Europe more difficult. While acceptance has been delayed in other places, the European Commission has launched the Digital Roadmap initiative, which aspires to 5G coverage in all major cities by 2030 [24].

4.6.11.4   Security in 5G Architecture

Through substantial recourse to cloud-based services, virtualization, network slicing, and additional upcoming skills, 5G adoption provides enormous performance improvements and application variety. However, these modifications have introduced new safety vulnerabilities into the 5G security architecture [25]:

- 5G security draws on earlier generations of cell phone technology, but the "rely on the framework" has increased with more stakeholders participating in delivery.
- The IoT and user propagation are exponentially expanding the number of terminations, with many of those network inputs no longer being overseen by human hands.
- In response to the threats, the 3GPP specifications outline enhanced 5G security features such as unified authentication to isolate verification from admission points and public key-based methods of encoding to limit the danger of metadata exploitation [26].
- As 5G essential nodes become progressively virtualized, continuous evaluation and tracking of security efficacy are crucial.
- Best practices for 5G network security include end-to-end tracking of the organization's architecture, plans, and apps.

Without question, 5G provides the exponential rapidity boost that customers have come to anticipate with every new era of cell phones, but speed is only the beginning [27]. The effects on businesses reaching from private transference to industry and agriculture are so profound that several have termed 5G the "next Industrial Revolution." The multifunctional 5G construction, with MEC, NFV huge MIMO, and a cloud-aligned, service-based fundamental style working in tandem to offer the next generation of services, lies at the heart of this paradigm shift [28]. The key organizers of the upcoming 5G migration will be 5G experimentation intended to support this architecture seed shift.

4.8   THE FUTURE OF 5G

Low latency is a crucial component of 5G technology, especially for self-driving vehicles and mission-critical applications. 5G networks can achieve latency of less than one ms. To achieve these speeds, 5G will use new radio millimeter waves to transfer data. It has significantly more bandwidth than lesser LTE channels and can process extremely high data speeds. In future generations, vehicles will be able to interconnect with traffic flow signals and adjacent objects, as well as other automobiles on the highway. Every millisecond matters in self-driving vehicles; findings must be complete in just a few seconds to avoid crashes and assure the security of passengers [29]. 5G will create tremendous opportunities:

- Expanded mobile broadband will deliver expanded capabilities to mobile network consumers and industrial users, boosting B2C, entertainment, and media, as well as extra service industries.
- URLLC will open up new use cases in mission-critical applications such as self-driving cars, virtual clinical procedures, and automation in the industry.
- The new mMTC will allow participants in industries such as manufacturing, utilities, and logistics to link huge numbers of devices with specialized connection requirements.
- Because of its extensive variety of technological and commercial prospects, the 5G paradigm brings together mobile network operators, equipment vendors, and different industry players more than any prior generation of technology.
- Because they will operate the 5G networks, network workers play a critical part in the 5G rollout. That will contribute the majority of the required capital expenditures and operating expenses.
- The equipment suppliers create a 5G network, as well as describe and launch 5G network products like RAN, core network, services, and phones [30].

New prospects with 5G bring industry players into the paradigm [31–33]. Many industries are participating as 5G will allow tremendous innovation on the product (e.g., linked cars) and productivity (e.g., industrial IoT) sides.

REFERENCES

1. www.rfpage.com/applications-5g-technology/
2. www.qualcomm.com/news/onq/2021/11/future-5g-growing-beyond-smartphone
3. https://rantcell.com/comparison-of-2g-3g-4g-5g.html
4. A Novel Credit Grounded Job Scheduling Algorithm for the Cloud Computing Environment Sundaravadivazhagan, B., Malathi, V., Kavitha, V. 5th International Conference on Inventive Computation Technologies, ICICT 2022 — Proceedings, 2022, pp. 912–919.
5. www.uctel.co.uk/blog/whats-the-difference-between-3g-4g-and-5g
6. www.ericsson.com/en/core-network/5g-core#:~:text=The%20new%205GC%20architecture%20is,Application%20Programming%20Interfaces%20(API).
7. Energy Aware Load Balancing Algorithm for Upgraded Effectiveness in Green Cloud Computing Malathi, V., Kavitha, V. 2022, 209, pp. 247–260.
8. 5G Core Network-Telecommunication Engineering Centre, New Delhi.
9. www.viavisolutions.com/en-us/what-5g-architecture#:~:text=5G%20creates%20a%20dynamic%2C%20coherent,station%20proximity%20or%20complex%20infrastructure.

10. Tobit regressive based gaussian independence bayes map reduce classifier on data warehouse for predictive analytics, Sivakkolundu, R., Kavitha, V., International Journal of Innovative Technology and Exploring Engineering, 2019, 8(12), pp. 4269–4280

11. Position Paper 5G Applications-Roland Berger, Huawei.

12. www.techtarget.com/whatis/definition/network-slicing.

13. www.4g.co.uk/what-is-4g/

14. www.thalesgroup.com/en/markets/digital-identity-and-security/technology/lte.

15. Nwazor NO, Ugah V. "Device-To-Device (D2D) data communications in 5g networks". International Journal of Ad-vances in Engineering and Management (IJAEM) IV. 1 (2022): 1151–1154.

16. Filali A, et al. "Multi-access edge computing: A survey". IEEEAccess 8 (2020): 197017–197046.

17. Bhamare D, et al. "A survey on service function chaining". Journal of Network and Computer Applications 75 (2016): 138–155.

18. Kamran R, et al. "A survey on multicast broadcast services in5G and beyond". (2022).

19. Kasim AN. "A survey mobility management in 5G networks".arXiv preprint arXiv:2006.15598, (2020).

20. Alotaibi S. "Key challenges of mobility management and handover process". International Journal of Computer Scienceand Network Security 22.4 (2022): 139–146.

21. Priscoli FD, et al. "Capacity-constrained Wardrop equilibria and application to multi-connectivity in 5G networks". Journal of the Franklin Institute 358.17 (2021): 9364–9384.

22. Lisi F, et al. "Multi-connectivity in 5G terrestrial-SatelliteNetworks: The 5G-ALL-STAR Solution". arXiv preprint arX -iv:2004.00368 (2020).

23. Parvez A., et al. "A survey on low latency towards 5G: RAN, core network and caching solutions". IEEE Communications Surveys and Tutorials 20.4 (2018): 3098–3130.

24. Kholidy HA. "Multi-layer attack graph analysis in the5G edge network using a dynamic hexagonal fuzzy method". Sensors 22 (2022): 9.

25. Kholidy HA, et al. "5G core security in edge networks: A vulnerability assessment approach". The 26 th IEEE Sympo-sium on Computers and Communications (IEEE ISCC 2021), Athens, Greece, September 5–8, (2021). A Survey Study For the 5G Emerging Technologies

26. *A Survey Study For the 5G Emerging Technologies*. Available from: www.researchgate.net/publication/369768507_A_Survey_Study_For_the_5G_Emerging_Technologies [accessed Jun 12 2023].

27. Jaber M, Imran MA, Tafazolli R, et al. 5G Backhaul challenges and emerging research directions: A survey [J]. IEEE Access, 4 (2017): 1743–1766.

28. Martrat J, Bernardos CJ, Chen T, et al. Vision on Software Networks and 5G. (2017).

29. Sezer S, Scott-Hayward S, Chouhan P, et al. Are we ready for SDN? Implementation challenges for software-defined networks[J]. IEEE Communications Magazine, 51.7 (2013): 36–43.

30. Jain A, Lopez-Aguilera E, Demirkol I. Improved Handover Signaling for 5G Networks[C]//2018 IEEE 29th Annual International Symposium on Personal, Indoor and Mobile Radio Communications (PIMRC). IEEE, 2018.

31. www.celona.io/5g-lan/5g-spectrum

32. www.gsma.com/spectrum/5g-spectrum-guide/

33. www.viavisolutions.com/en-us/what-5g-architecture

Chapter 5

5G—Further Evolution

*Avanthica Sri M M, Chandrakumar T, Gautam
Srivastava, and Subrata Chowdhury*

5.1 INTRODUCTION

Consumer expectations are playing a vital role in shaping the expansion of mobile broadband services. The surge in traffic, which is projected to escalate by 10 to 100 times from 2020 to 2030, coupled with the proliferation of devices and services and the desire for enhanced affordability and customer experience will all call for innovative approaches to address these challenges. According to a search, there are now 6.2 billion potential mobile customers worldwide, up from 4.4 billion in 2013 and 5.4 billion in 2017. This development is expected to continue through the end of 2023 [1]. By the year 2025, it is anticipated that the total count of interconnected devices on the Internet will surge to 50 billion.

The deployment of 5G in today's smart cities has successfully ushered in a digital era dominated by intelligent wireless sensor nodes [2]. As of the year 2024, we witness a remarkable shift, with over 60% of the global population now enjoying the benefits of widespread 5G coverage. This substantial transition is mirrored in the distribution of network technologies, where 2G now occupies a mere 34.5%, 3G claims 38.5%, and 4G contributes 27% of the network space, as of the year 2020 [3]. These changes underscore the persistent global initiatives to enhance connectivity, driven by a proliferation of devices and an escalating demand for faster data speeds. The integration of 5G seamlessly intertwines people, activities, data, applications, transportation systems, and urban landscapes through a sophisticated network of adaptive communication, marking a significant milestone in our digital evolution.

The implementation of 5G technology in urban areas will enhance traffic management systems through its ability to accommodate a significant volume of Internet of Things (IoT) connections to traffic lights, cameras, and traffic sensors. This paper [4] proposes a new paradigm, 5G intelligent IoT, for intelligently processing massive data and optimizing communication channels. This article [5] provides an in-depth examination of cutting-edge concepts that use 5G-enabled IoT as a backbone for blockchain-based industrial automation in applications such as smart homes, smart cities, smart agriculture, health care 4.0, supply chain management, and autonomous cars. This paper [6] presents a hierarchical design for protecting 5G-enabled IoT networks, as well as a security model

DOI: 10.1201/9781003306290-5

for predicting and detecting false data injection attacks and distributed denial of service. This article [7] discusses some of the significant opportunities that these technologies present, as well as the obstacles that the microwave and wireless communication communities face. This article [8] describes the future 6G wireless communication vision and network architecture. The networks facilitate the rapid transfer of vast quantities of data, reliably connecting a large number of devices, and efficiently processing extremely large data volumes with minimal latency. Everything is new as a result of generational changes in technology.

5.2 5G CELLULAR SYSTEM

The fifth-generation mobile technology, 5G, is the most recent advancement in the LTE mobile broadband networks, which began their worldwide deployment in 2019 and are projected to provide connectivity to over 1.7 billion individuals by the conclusion of 2025. When comparing 5G networks with their predecessor, 4G, noticeable improvements can be observed in terms of reduced latency, increased capacity, and enhanced bandwidth. These far-reaching network advancements will significantly impact various aspects of people's lives, work, and recreational activities worldwide. The 5G cellular system represents the most recent iteration of wireless technology, offering accelerated data speeds, reduced latency, and enhanced connectivity that is characterized by greater reliability. Figure 5.1 displays the important features of 5G cellular systems that make 5G stand out from the previous generation.

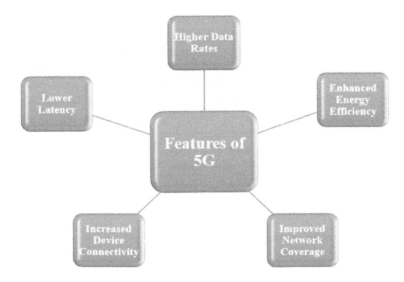

Figure 5.1 Features of a 5G Cellular System.

The following are some of the important aspects of the 5G cellular system:

- **Higher data rates:** The 5G wireless technology has been engineered to deliver data speeds reaching up to 10 Gbps, a significant improvement compared with the speeds offered by earlier generations of wireless technology.
- **Lower latency:** Compared with earlier generations, 5G demonstrates lower latency, resulting in a decreased time gap between the transmission of a request and the receipt of a corresponding response.
- **Improved network coverage:** 5G has the ability to offer better coverage in more locations because it employs a wider spectrum of frequencies than older generations did.
- **Increased device connectivity:** The design of 5G technology prioritizes the ability to fit a significantly number of devices compared to previous generations. This aspect holds great significance considering the exponential growth of interconnected devices within the IoT.
- **Enhanced energy efficiency:** One notable advantage of 5G is its improved energy efficiency compared with earlier generations. This design feature allows 5G to consume less power, resulting in reduced carbon emissions and a positive environmental impact.

5.2.1   Reduced Latency

Reducing latency has been a consistent objective for each successive wireless generation, and the upcoming 5G networks will achieve even reduced latency compared with 4G. With 5G, the round-trip data transfer will require fewer than five milliseconds, surpassing the speed of human visual processing. This progress allows for the remote control of equipment with nearly real-time responsiveness.

The reduced latency offered by 5G unlocks new possibilities for machine-to-machine communication free from the constraints of human reaction time. In the context of remote applications utilizing 5G and IoT, the limiting factor will shift to human reaction speed, implying that the technology will enable responsiveness beyond human capabilities. In addition to its benefits in sectors like agriculture, logistics, and manufacturing, the gaming community is eagerly awaiting the arrival of 5G. The unique combination of high-speed data transmission and minimal lag in 5G networks makes them ideal for supporting augmented reality (AR) and virtual reality (VR) applications. As connectivity advances, these immersive experiences are expected to gain popularity due to their seamless and captivating nature.

5.2.2   Enhanced Capacity

With its significantly increased capacity, 5G has the potential to greatly support the scaling of technology initiatives for businesses. The extensive capacity of 5G, which can be up to 1000 times bigger than 4G, creates an ideal environment for the development of IoT applications. The combination of IoT and 5G has the power to transform the utilization of the Internet and wireless networks as a whole. The seamless communication capabilities of 5G networks facilitate the integration of a vast number of devices, paving the way for the emergence of new applications and use cases across diverse domains including factories, cities, farms, homes, and schools. For example, consider the potential of 5G technology in incorporating thousands of sensors across multiple machines, revolutionizing supply chain management processes. With 5G-enabled connectivity, these sensors can facilitate real-time monitoring of inventory levels, ensuring timely delivery of materials through efficient just-in-time practices.

The future of 5G holds significant advancements for smart homes and cities. With the proliferation of connected devices, the integration of artificial intelligence (AI) will reach new frontiers through edge computing. This progress will enable personalized energy-saving recommendations for homes, maximizing their environmental impact. The increased network capacity of 5G will be instrumental in facilitating these applications, leading to a significant impact on various societal aspects. Industries will benefit from enhanced capabilities in automation and efficiency, while smart city infrastructure, including traffic lights and other systems, can be seamlessly integrated into the network for improved functionality. The expanded capacity of 5G networks will pave the way for a wide range of transformative applications that have the potential to revolutionize multiple sectors and contribute to the development of smarter and more interconnected communities.

5.2.3 Improved Bandwidth

Compared with 4G LTE, 5G networks offer the potential for greater data transmission volumes thanks to their faster speeds and enhanced network capacities. The innovative architecture of 5G networks enables traffic optimization and effective handling of surges in data usage. With its ability to provide continuous and seamless connectivity, 5G technology has the capacity to transform the way large crowds engage with live events, offering an enhanced and immersive experience for spectators. For instance, this advancement presents exciting opportunities for sports enthusiasts to stream their live experiences from any location within a stadium, addressing the issue of maintaining uninterrupted connectivity in densely populated venues.

The increasing bandwidth made available through 5G technology will also have significant impacts on business and industry, particularly in the realm of big data. Companies are now inundated with a wealth of data from customers, teams, and suppliers, surpassing their capacity to effectively analyze and extract insights. However, with the integration of big data analytics and 5G connectivity, companies can harness the power to unlock the potential of their vast amounts of data and derive actionable insights that drive informed decision-making and business success.

5.2.3.1 5G Technology Utilizes Three Primary Bands to Operate Effectively

- **Low band spectrum:** This band offers extensive Internet coverage and enables transfer of transfer at respectable rates, although its biggest speed is limited to 100 Mbps.
- **Mid band spectrum:** This spectrum provides faster speeds than the lower band, but it has certain constraints on signal penetration and also coverage area.
- **High band spectrum:** Also known as millimeter wave, this band boasts the highest speeds among all the spectrum bands. However, its range and signal intensity are significantly constrained, resulting in limited coverage areas and requiring closer proximity to the transmission source for optimal performance.

5.2.4 Speed

The 5G signal range is a drawback. Because it utilizes high-frequency radio waves, 5G technology requires smaller geographic cells, which, in turn, necessitate additional towers. This shift to smaller cells increases the deployment costs associated with 5G networks. To accommodate the diverse needs of different applications, 5G will be deployed across three frequency bands: low band that spans between 600 and 700 megahertz, mid band that spans between 2.5 and 3.7 gigahertz (GHz), and high band spans between 25 and 39 GHz. Each frequency band offers specific advantages and characteristics, enabling a range of use cases and ensuring efficient network performance in various scenarios.

Furthermore, 5G technology will significantly reduce latency, which refers to the delay in sending and receiving requests between a cell phone or connected device and servers. With 5G, the latency will be greatly minimized, leading to almost instantaneous communication. Moreover, 5G will enhance and expedite connections to cloud computing platforms, namely Amazon Web Services and Microsoft Azure. This integration with cloud services will enable faster and more efficient data processing, storage, and access, unlocking new possibilities for applications and services that rely on cloud computing capabilities.

5.3 WORKING TECHNOLOGIES

To deliver high-speed wireless communication, 5G networks leverage a combination of novel and existing technologies, including advanced antennas, beamforming, and network slicing. These technological advancements empower 5G networks to deliver reduced latency and enhanced data speeds compared with preceding generations of wireless technology. One of the key technologies employed in 5G networks is millimeter wave (mmWave) technology, which utilizes high-frequency radio waves to transmit data. This advanced technology plays a fundamental role in facilitating the high-speed data transfer capabilities of 5G networks. By harnessing the power of mmWave frequencies, 5G networks can support faster data transmission rates and enable the efficient delivery of large volumes of data across wireless networks.

The 5G networks also employ a method known as beamforming, which allows the network to focus radio waves in a precise direction, enhancing network efficiency and dependability, and use carrier aggregation to enable high-speed wireless communication, which allows many wireless channels to be combined into a single, faster link. Another important technology employed in 5G networks is massive multiple input, multiple output (MIMO) antenna arrays to enhance network capacity and efficiency; by combining a large number of antennas, 5G networks can support simultaneous transmission and reception of multiple data streams, ensuing increased data throughput and better network performance. Massive MIMO enables higher spectral efficiency, improved coverage, and better overall network performance, making it a key technology for enhancing the capabilities of 5G networks. Network slicing also enables customization and optimization of network resources for different use cases, providing tailored connectivity solutions for various industries and applications. By efficiently allocating network resources, 5G networks can deliver optimal performance and meet the diverse needs of different applications and consumers.

Signals used by 5G networks operate on new radio frequencies, requiring the upgrade of radios and other hardware components on cell towers. While 5G is currently limited to smartphones, there is anticipation that this technology will revolutionize home Internet usage. The expansion of 5G in residential settings has the capacity to transform how we connect and utilize the Internet within our homes. Once businesses can provide Internet to homes without having to lay down connections, traditional Internet service providers will face serious threats. Consumers must own 5G-capable devices in order to connect to and utilize a 5G network.

5.4 GLOBAL SCENARIO

By 2025, it is projected that commercial 5G networks will account for approximately 12.5% of global mobile connections, which translates to around 1.1 billion connections. This significant adoption of 5G technology is expected to generate substantial revenue for operators, reaching up to $1.3 trillion. The widespread implementation of 5G networks will not only revolutionize connectivity but also present lucrative opportunities for operators to capitalize on the increasing demand for advanced mobile services. The adoption and deployment of 5G networks are underway in numerous countries worldwide, with certain nations emerging as leaders in this technological advancement.

Notably, countries such as the United States, South Korea, and China have made substantial strides in deploying 5G networks. These countries have been proactive in embracing the potential of 5G technology and have taken significant steps to establish robust and widespread coverage. As a result, they are at the forefront of the global 5G landscape, setting the stage for transformative advancements in connectivity and communication.

- In the United States, several telecommunications companies have already launched 5G networks in select cities, and coverage is expected to expand rapidly over the next few years. The US government has also allocated significant resources to guide the development and expansion of 5G networks. As of now, 5G technology has been successfully established in 50 cities across the United States.
- China is also making significant investments in 5G technology, with several major Chinese telecommunications companies launching 5G networks in select cities. In the near future, China is anticipated to play a significant role in the global 5G market. Japan and China have begun trialing 5G mobile services.
- South Korea was one of the first countries to launch 5G networks, and it has made significant progress in terms of adoption. As of 2021, more than 10 million South Koreans were using 5G networks, and the country has plans to expand coverage to more areas. South Korea has expanded 5G service to 85 cities.
- In Europe, several countries have launched 5G networks, including the United Kingdom, Germany, and Spain. However, coverage is still limited in many areas, and there are issues around the cost and feasibility of deploying 5G networks in more rural areas.
- Starting from October 1, 2022, the initial rollout of 5G services in India began, covering 13 major cities: Chennai, Delhi, Kolkata, Mumbai, Bengaluru, Gurugram, Chandigarh, Hyderabad, Jamnagar, Lucknow, Ahmedabad, Gandhinagar, and Pune, but the plan is to expand 5G coverage nationwide. In addition, major smartphone manufacturers such as Apple, Google, and Samsung have committed to updating their recent smartphone models to support 5G technology, ensuring compatibility with the upcoming 5G services in India. This demonstrates their dedication to providing users with the benefits and capabilities of 5G connectivity on their devices.

The 5G cellular system is being rolled out in phases, with some cities and regions already having access to 5G networks. However, it will likely be several years before 5G networks are widely available in all areas. In the foreseeable future, the global market for 5G technology is poised for substantial growth, with numerous industry experts projecting that 5G networks will emerge as the prevailing wireless technology in the next decade.

5.5 5G USE CASES

In terms of its actual applications, 5G is expected to transform or change various industries like transportation, health care, and manufacturing. For example, 5G networks could enable remote surgeries in rural, medically understaffed areas or help to reduce traffic congestion by providing real-time traffic updates to drivers.

- **Enhanced mobile broadband:** 5G can provide faster download and upload rates, greater capacity, and lower latency, which enhances the user experience.
- **The IoT:** This next-generation wireless technology enables the seamless integration of a multitude of devices, ranging from smart homes and cities to industrial automation systems, and 5G technology is poised to revolutionize it with its ability to manage a huge volume of connected devices, lower latency, and improved reliability.
- **VR and AR:** 5G offers the high-speed, low-latency connectivity required for immersive AR and VR experiences in entertainment, education, and other applications.
- **Autonomous Vehicles:** 5G can deliver the low-latency, high-reliability connectivity required for autonomous vehicles to communicate with one another and with infrastructure, enhancing safety and enabling new transportation services.
- **Remote Health Care:** 5G can deliver the high-speed, low-latency connectivity required for remote health care services like telemedicine, remote monitoring, and remote surgery, which will expand access to health care services and lower costs.

5.6 BENEFITS OF 5G

- The higher bandwidth that can accommodate multiple connected devices is 5G's main advantage over 4G. It is a hundred times beyond the maximum speed of 4G and offers a high band spectrum with a potential of 20 Gbps. It is quick enough to install a movie within seconds with good resolution.
- In addition to making services such as health care and mobile banking more accessible, 5G networks may also open up an exponential increase in chances for jobless people to find meaningful employment.
- Enabling communities and residents to experience the socioeconomic conveniences and advantages of a highly advanced, data-intensive digital economy would become a reality with the implementation of such technology.
- It employs a millimeter-wave spectrum, allowing for the utilization of approximately one million devices per square kilometer in a given area.
- It makes use of new digital technology that boosts capacity, speed, and coverage.

5.7   LIMITATIONS OF 5G

- **Critical infrastructure:** In order to accommodate the necessities of 5G technology, fundamental modifications will need to be made to existing communication system designs. One of the primary challenges faced by 5G is its limited data transmission range, which necessitates augmenting current infrastructure, installing additional cell towers and deploying fiber optic cables to assist the elevated data traffic because these enhancements come at a considerable cost. Furthermore, the implementation can be time-consuming and complex, as it involves integrating new technologies with the existing network infrastructure.

- **Financial liability:** Upgrading from 4G to 5G technology requires consumers to invest in devices and equipment that are compatible with the new technology.

- **Lack of sufficient capital:** Competent telecom companies like Vodafone and Airtel are facing delays in the allocation of 5G spectrum due to a shortage of adequate capital. The expansion of 5G networks requires significant investment in acquiring the necessary spectrum licenses and upgrading infrastructure.

- **Delayed implementation:** Several countries in the Asia-Pacific region, such as India, Bangladesh, and Indonesia, have experienced a slower adoption of 5G technology than other regions. This delayed adoption may potentially lead to lower revenue generation from 5G services in these countries.

- **Limited government subsidies:** The likelihood of receiving government subsidies for 5G infrastructure development is low, primarily due to the historical trend of governments setting greater reserve prices for spectrum sales. Despite persistent budget deficits, governments have been reluctant to provide substantial financial support in the form of subsidies for the deployment of 5G networks. This lack of subsidies can pose financial challenges for telecom operators and hinder the widespread adoption of 5G technology.

- **Digital divide:** 5G won't immediately close the gap between urban and rural areas in terms of access to the Internet; in fact, it might widen it.

- **Niche service:** Unlike the widespread distribution of services seen in 3G and 4G networks, 5G is expected to cater to specific niches and industries. Rather than being rolled out quickly across the board, the implementation of 5G will be more targeted and concentrated in specific sectors.

- **Persisting network issues:** Consumers still encounter fundamental network problems, including call dropouts and inconsistent broadband connections. The existing 4G networks often disrupt Internet services, causing inconvenience and frustration for consumers.

- **Interference:** There are fears that the introduction of 5G mobile services will interfere with aero-plane navigation systems. Air India suspended eight flights to US destinations in January 2022.

5.8   5G IN IOT

The Internet of Things is a concept that involves integrated physical devices, home appliances, vehicles, and other objects provided with software, sensors, and connectivity to enable data exchange. IoT has the capacity to transform our daily lives and professional environments by facilitating communication between devices and individuals, leading to a more interconnected and streamlined world. The applications of IoT span across various sectors, including smart homes, cities, health care, and industrial automation, offering improved efficiency and convenience.

The new 5G in IoT allows for quicker, more reliable, and secure connectivity between devices. Because of the reduced latency of 5G networks, real-time data processing is possible, which is critical for many IoT applications and enables the design of innovative applications and services that were previously unimaginable. The expansive proportions of 5G networks allow for the seamless connection of a multitude of devices, which is crucial for IoT applications involving numerous sensors and devices. Moreover, 5G networks are energy-efficient compared with previous wireless network generations, making them ideal for IoT devices that require prolonged operation on a single battery charge. The authors of [9] detail a network model and a threat model that are required for a 5G-enabled IoT communication environment including the intricacies of the security needs and probable assaults in this communication environment are expanded.

The introduction of 5G technology will lead to a vital increase in data transfer speeds. According to reports, 5G is expected to be 10 times faster than LTE networks. This enhanced speed enables IoT devices to transfer and share data at a much quicker rate than ever before. Blockchain's integration with 5G and IoT has potential financial benefits but still requires critical insights into specific application areas, scalability, privacy concerns, and performance. The purpose of this paper [10] was to expound on and emphasize the main aspects of using Blockchain for 5G and IoT. This work [11] used blockchain distributed networks to overcome network scalability issues and the Raft consensus algorithm to boost blockchain throughput. Additionally, 5G-enabled smart meters equipped with low-cost IoT sensors and connections will track energy usage and donate to energy consumption reduction efforts. Under the 5G standard for massive IoT, it is estimated that up to one million connected devices can be supported within an area of approximately 0.38 square miles (1 square km).

5.8.1   Massive Device Connectivity

Supporting a vast number of connected devices at once is one of 5G's primary advantages in the IoT. This scalability is crucial for the development of IoT applications because it enables the smooth connection and communication of a wide variety of devices, including wearables, smart appliances, industrial machinery, and sensors.

5.8.2 Enhanced Coverage and Dependability

The increased coverage and dependability of 5G compared with earlier cellular networks make it highly suited for IoT installations. The frequency ranges used by 5G networks include low-band frequencies, which provide more coverage and better obstacle penetration. This makes it possible for IoT devices to have dependable communication even in difficult settings like subterranean facilities or isolated places.

5.8.3 Enhanced Security and Privacy

To meet the particular difficulties of IoT installations, 5G integrates advanced security capabilities. It offers increased privacy safeguards, stronger authentication procedures, and encryption algorithms for IoT communications. Strong security measures are essential for stopping unauthorized access, data breaches, and other cybersecurity concerns due to the prevalence of IoT devices and the sensitive data they produce.

5.8.4 Transformative IoT Applications

By combining 5G and IoT, transformative applications across sectors are made possible. Several prominent use cases include:

- **Smart Cities:** 5G-enabled IoT can support effective city management, including smart grids for electricity and transportation, intelligent systems for observing the environment, and systems for public safety.
- **Industrial Automation:** 5G offers predictive maintenance, real-time monitoring and management of industrial processes, and seamless machine-to-machine connection, all of which boost production and efficiency.
- **Health Care:** 5G IoT devices can enable telemedicine, real-time health tracking, remote patient monitoring, and enhanced emergency response systems.
- **Agriculture:** 5G-enabled IoT technologies can boost crop yields and resource efficiency through precision agriculture, remote monitoring, soil and crop health monitoring, and automated irrigation systems.

Figure 5.2 depicts the use cases of 5G-enabled IoT, which provides unrivalled connectivity, speed, responsiveness, and scalability, enabling the development of novel applications across industries. It has the potential to open up new opportunities for automation, real-time analytics, and disruptive IoT deployments.

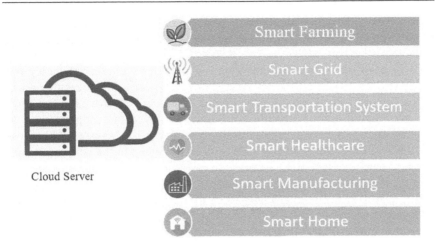

Figure 5.2 5G-Enabled IoT Applications.

5.9 5G AND BEYOND

With immense capacity (upwards of 1000 times) and connectivity (billions of users and machines), the next-generation mobile wireless communication networks are expected to enable ground-breaking mobile applications. These 5G applications require high-quality low-latency visual, tactile, and auditory telepresence. This paper [12] provides a high-level overview of 5G NR's new features and important performance indicators. This work thoroughly addresses the concerns of higher modulation scheme adaptability and inter-RAT handover synchronization and presents a next-generation wireless communication architecture.

Next-generation mobile communications have started using frequencies at millimeter wavelengths (30–300 GHz), where high-power transmitters with massive MIMO antennas can recover from severe propagation loss via beamforming; more antennas produce higher transmit power. The utilization of mmWave spectrum to ease the spectrum squeeze in existing frequency bands, massive MIMO antenna arrays to boost simultaneous transmission capacity, and ultra-dense networks to enable short-range, high-speed data transfer are all being investigated by the industry.

Scientists and engineers are actively engaged in developing technologies for future networks that will surpass the capabilities of 5G, despite 5G networks being in their initial stages of deployment. These ongoing efforts reflect a forward-looking approach to continually push the boundaries of wireless communication technology. One of the main objectives of beyond 5G is to make device connection even faster and more dependable, such as with terahertz (THz) communication, which operates at frequencies higher than those of present-day wireless networks. Data rates that are several orders of magnitude faster than those currently feasible with 5G could be made possible through THz communication. Enabling more device connections is a key focus of future wireless technologies beyond 5G.

The goal of next-generation wireless technologies includes enhancing the energy efficiency of wireless networks, particularly for IoT devices that operate on limited battery power. Innovations like energy harvesting and low-power communication protocols are being developed and implemented to optimize energy consumption and prolong battery life. This paper [13] presents a 6G vision that could serve as a research roadmap in the post-5G future. Its authors believe that human-centric mobile communications will continue to be the most important use of 6G and that the 6G network should be human-centric as well and also investigates topics that could impede 6G research and deployment.

Another important technology for beyond 5G is AI. AI has the capacity to transform wireless networks by optimizing their performance and enhancing resource management. By leveraging AI algorithms and techniques, wireless networks can achieve more systematic utilization of the applicable spectrum, guiding to improved network capacity, coverage, and overall performance. AI can analyze network data in real time, identify patterns, and make intelligent decisions to dynamically allocate resources, prioritize traffic, and adapt to changing network conditions. AI is currently being used in new applications such as intelligent traffic management and predictive maintenance for industrial equipment.

5.10   WHEN CAN WE EXPECT THE ADVENT OF 6G?

As the next generation of wireless technology, 6G is anticipated to offer blazing-fast speeds, with data transfer rates reaching to 96 Gbps. This speed is approximately 10 times faster than what is currently achievable with 5G networks. The technique improves spectrum sharing, lowers latency, and increases capacity by making greater use of the THz spectrum and distributed RAN. Although preliminary tests are being conducted in China, South Korea, and Japan, it is anticipated that the technology won't be made widely available until the 2030s. It is believed that 6G will be more than just increased spectrum in high-frequency bands; rather, it will represent a convergence of forthcoming technological developments driven by interesting, underlying services [14].

5.11   WHAT IS 6G?

The upcoming iteration of wireless network technology is 6G; it represents the next phase of advancement in wireless communication systems. Despite 5G networks being in their early stages, researchers and engineers are already focusing on the development of 6G networks, aiming for even greater speed, reliability, and efficiency. The purposes of [15] and [16] were to provide a full review of the 6G system in terms of use cases, drivers, key performance indicators, needs, supporting technologies and architecture. Some features of 6G networks include significantly faster data rates, reduced latency, and optimized spectrum utilization. The implementation of 6G networks will unlock new possibilities for advanced applications and services, including enhanced virtual and augmented reality experiences, autonomous vehicles, and smart cities. However, the widespread availability of 6G networks is still years away as the technology continues to evolve and mature. Figure 5.3 displays the 6G features that will be available in the near future.

While 6G research is still in its initial stages, its potential features and technologies have been anticipated.

5.11.1   Terahertz Frequencies

Thz frequencies, which are significantly higher than the radio frequencies utilized in 5G, are expected to be explored in 6G. Extremely high data rates, possibly reaching tens or even hundreds of gigabits per second, are possible at THz frequencies.

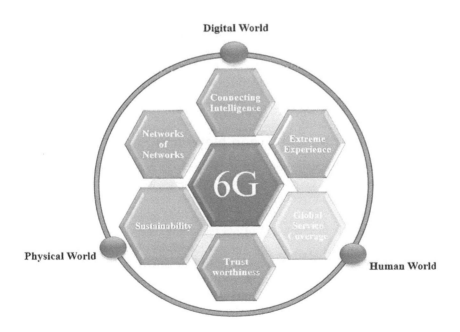

Figure 5.3 6G Features.

5.11.2   Holographic Beamforming

Holographic beamforming techniques could replace traditional beamforming, which directs wireless signals in a certain direction. This would make it possible to manipulate wireless signals more precisely, resulting in improved coverage and increased capacity.

5.11.3   AI-Driven Networking

In 6G networks, AI is probably going to be quite important in making networks operate more effectively and autonomously. AI algorithms can optimize how network resources are distributed, improve security, and enable intelligent network management.

5.11.4   Integrated Satellite Networks

The upcoming 6G will combine terrestrial and satellite networks, providing connectivity and global coverage. With this connection, numerous IoT applications, moving vehicles, and remote locations will all have seamless connectivity.

5.11.5   Quantum Communications

An emerging field that harnesses the principles of quantum physics to ensure highly secure and impervious data transfer is quantum communications. In the context of 6G, there is a possibility of incorporating quantum communication technologies to establish exceptionally secure communication channels. These advancements could offer unprecedented levels of data protection, making it extremely difficult for unauthorized entities to intercept or tamper with transmitted information.

5.11.6   Sustainable and Green Networks

Energy efficiency and sustainability are expected to be key considerations in 6G design. Research is being conducted to develop energy-efficient technologies and network architectures that reduce carbon footprint and ensure environmentally friendly operations.

The progress and implementation of 6G technology are expected to take several years before becoming a reality. Various standards organizations, researchers, and industry players are currently engaged in defining the specifications and potential of 6G networks. While the specific timeline may vary, it is projected that the initial commercial rollout of 6G networks will likely commence in the 2030s. This extended timeframe allows for comprehensive research, development, and standardization efforts to ensure that 6G meets the evolving needs of future wireless communication.

REFERENCES

[1] Hossain, M. F., Mahin, A. U., Debnath, T., Mosharrof, F. B., & Islam, K. Z. (2019). Recent research in cloud radio access network (C-RAN) for 5G cellular systems-A survey. Journal of Network and Computer Applications, 139, 31–48.

[2] Zhao, X., Askari, H., & Chen, J. (2021). Nanogenerators for smart cities in the era of 5G and Internet of Things. Joule, 5(6), 1391–1431.

[3] Nadeem, L., Azam, M. A., Amin, Y., Al-Ghamdi, M. A., Chai, K. K., Khan, M. F. N., & Khan, M. A. (2021). Integration of D2D, network slicing, and MEC in 5G cellular networks: Survey and challenges. IEEE Access, 9, 37590–37612.

[4] Wang, D., Chen, D., Song, B., Guizani, N., Yu, X., & Du, X. (2018). From IoT to 5G I-IoT: The next generation IoT-based intelligent algorithms and 5G technologies. IEEE Communications Magazine, 56(10), 114–120.

[5] Mistry, I., Tanwar, S., Tyagi, S., & Kumar, N. (2020). Blockchain for 5G-enabled IoT for industrial automation: A systematic review, solutions, and challenges. Mechanical Systems and Signal Processing, 135, 106382.

[6] Moudoud, H., Khoukhi, L., & Cherkaoui, S. (2020). Prediction and detection of fdia and ddos attacks in 5g enabled iot. IEEE Network, 35(2), 194–201.

[7] Gustavsson, U., Frenger, P., Fager, C., Eriksson, T., Zirath, H., Dielacher, F., . . . & Carvalho, N. B. (2021). Implementation challenges and opportunities in beyond-5G and 6G communication. IEEE Journal of Microwaves, 1(1), 86–100.

[8] Chowdhury, M. Z., Shahjalal, M., Ahmed, S., & Jang, Y. M. (2020). 6G wireless communication systems: Applications, requirements, technologies, challenges, and research directions. IEEE Open Journal of the Communications Society, 1, 957–975.

[9] Wazid, M., Das, A. K., Shetty, S., Gope, P., & Rodrigues, J. J. (2020). Security in 5G-enabled internet of things communication: Issues, challenges, and future research roadmap. IEEE Access, 9, 4466–4489.

[10] Hewa, T. M., Kalla, A., Nag, A., Ylianttila, M. E., & Liyanage, M. (2020, October). Blockchain for 5G and IoT: Opportunities and challenges. In 2020 IEEE Eighth International Conference on Communications and Networking (ComNet) (pp. 1–8). IEEE.

[11] Dhar Dwivedi, A., Singh, R., Kaushik, K., Rao Mukkamala, R., & Alnumay, W. S. (2021). Blockchain and artificial intelligence for 5G-enabled Internet of Things: Challenges, opportunities, and solutions. Transactions on Emerging Telecommunications Technologies, e4329.

[12] Dogra, A., Jha, R. K., & Jain, S. (2020). A survey on beyond 5G network with the advent of 6G: Architecture and emerging technologies. IEEE Access, 9, 67512–67547.

[13] Dang, S., Amin, O., Shihada, B., & Alouini, M. S. (2020). What should 6G be?. Nature Electronics, 3(1), 20–29.

[14] Saad, W., Bennis, M., & Chen, M. (2019). A vision of 6G wireless systems: Applications, trends, technologies, and open research problems. IEEE Network, 34(3), 134–142.

[15] Jiang, W., Han, B., Habibi, M. A., & Schotten, H. D. (2021). The road towards 6G: A comprehensive survey. IEEE Open Journal of the Communications Society, 2, 334–366.

[16] Wang, C. X., You, X., Gao, X., Zhu, X., Li, Z., Zhang, C., . . . & Hanzo, L. (2023). On the road to 6G: Visions, requirements, key technologies and testbeds. IEEE Communications Surveys & Tutorials.

Chapter 6

Security and Privacy

Om Prakash, Saumya Das, Sreejith L Das, Satheesh Kumar Jaganathan and T Somassoundaram

6.1 PHYSICAL SECURITY USING MACHINE LEARNING

Physical security refers to security measures meant to shield people and things from danger and destruction and prevent unauthorized access to buildings, tools, and resources. Many interconnected levels of defense are necessary for physical security, including closed-circuit TV monitoring, access control, perimeter incursion detection, deterrent systems, protective barriers, locks, security personnel, and other systems.

Recent years have seen tremendous advancements in machine learning (ML) techniques. They are widely employed in various industries, including intelligent health care, self-driving automobiles, speech recognition, and natural language processing. In some cases, such as traditional image classification, ML is even more accurate than people are. Additionally, ML has enabled new security features and capabilities in security detection scenarios like spam filtering and harmful program identification. However, recent research reveals that ML models themselves are vulnerable to numerous security risks:

- Training data contamination might make models inaccurate or result in other error-specific or generic attack goals.
- A carefully constructed backdoor in the training data can have disastrous effects on a system.
- The model can fail if an adequately constructed perturbation in the test input (adversarial examples) occurs.
- Models can be stolen. Attacks using model inversion and membership inference can retrieve private training data or take control of the model parameters.

These security risks significantly impact ML systems, especially in applications crucial for security and safety, including autonomous driving, intelligent safety, innovative health care, etc.

ML security has received much attention recently [1, 2]. Since the danger of adversarial instances in deep learning (DL) algorithms was revealed, there has

DOI: 10.1201/9781003306290-6

been much research on the security of these systems. Spam detection, PDF virus identification, intrusion detection, and other tasks were investigated using so-called adversarial ML on non-deep ML algorithms; the majority of early assaults are evasion assaults, although some are known as poisoning assaults [3, 4, 5].

Overall, there have been few reviews on the subject of privacy and security problems in ML. In 2010, Barreno et al. assessed previous evasion attacks on non-DL systems [6] and provided an example of a spam filter. Akhtar and Mian [7] examined computer vision DL adversarial example assaults, emphasizing computer vision while discussing malevolent example attacks. Yuan et al. present a review of adversarial examples for DL [8]. Some investigators have listed damaging examples and examined defenses incorporating provably secure privacy-preserving ML algorithms with an emphasis on cryptographic primitives [8, 9].

Liu et al. examined the benefits and security risks of machine learning [10], and others concentrated on data security and security evaluation with a focus on systematizing ML security and privacy concerns [11]. These authors primarily analyzed the defenses with relation to robustness, accountability, and privacy while describing the assaults concerning three traditional security attributes, namely confidentiality, integrity, and availability [11].

6.2 ATTACK PREVENTION/DETECTION USING ML

One of the most significant risks to businesses and organizations worldwide is cybercrime. As cybercriminals become more skilled, conventional cybersecurity measures like firewalls and antivirus software are losing effectiveness. ML is a promising strategy for enhancing cyberattack detection and prevention.

Artificial intelligence (AI) is a subset of technology that enables computers to learn from data and develop without explicit programming. ML algorithms are practical tools for cybersecurity because they can analyze enormous amounts of data and spot trends. They can be trained on past data to learn how various attacks look in the context of attack detection and prevention. These algorithms can then use this information to recognize and stop similar attempts instantly. The two primary types of ML algorithms used in cybersecurity are supervised and unsupervised learning.

Supervised learning entails raining a model on labeled data, where the proper categorization of each data point is known. The machine then learns the traits of various types of attacks using this training data. The model can forecast the likelihood of an episode depending on the characteristics of new data submitted to it, for instance in detecting spam emails. The model is trained on labeled data that includes samples of both spam and non-spam emails.

Contrarily, in unsupervised learning, a model is trained on unlabeled data and must recognize patterns and structure in the data without labels; this strategy is helpful when dealing with novel attacks. Unsupervised learning can be used to group related data sets and spot odd patterns that can point to an assault. For instance, clustering techniques can categorize network traffic, with each category indicating a particular attack type.

ML algorithms can also be utilized in conjunction with existing security measures to enhance attack detection and prevention. For instance, traditional methods like firewalls and intrusion detection systems can be used to prevent or alert the activity. ML algorithms can also analyze log files from network devices to detect anomalous activity.

In conclusion, ML is a viable strategy for enhancing cybersecurity attack detection and prevention. Organizations can detect and stop cyberattacks in real-time using ML algorithms, but to provide adequate protection against cyber threats, it is crucial to remember that machine learning is not a panacea and must be used with other security measures.

6.3   EAVESDROPPING DETECTION USING ML

Every good thing has a downside, which is no different in the modern digital age. Today, although the Internet has improved our lives, it also presents a serious threat to data security, making hacks more likely. However, what precisely is a cyberattack?

Every purposeful and hostile attempt to get around an individual's or organization's information system constitutes a cyberattack. While most assaults aim to gain financial gain, some widespread attacks have as their primary objective the destruction or compromise of data. Eavesdropping is one such assault.

Eavesdropping is the practice of secretly listening in on other people's conversations. In hacking, it is a man-in-the-middle (MitM) attack that occurs when someone intercepts, alters, or deletes data transmitted between two devices in real time. Data examples include phone calls, instant messages, video chats, fax transmissions, and more. Attacks of this nature are also referred to as sniffing or spying. Your information, including passwords, credit card numbers, and other sensitive data, is easily stolen through these assaults as it is moved from one device to another. When a user connects to an unsecured or unencrypted network, sensitive company information is frequently sent to a coworker. Since the web is open, an attacker can take advantage of weakness and intercept the data using various techniques.

Attacks by eavesdroppers are frequently challenging to detect. A bug or listening device may not negatively impact the performance of devices and networks, in contrast to other types of cyber-attacks. There are two sorts of eavesdropping attacks: passive and aggressive. Active eavesdropping attacks include hackers entering the network and impersonating trusted connections; hackers can inject, change, or block transmissions during active attacks. MitM attacks are the most popular active eavesdropping attack. Most MitM attacks infiltrate systems via malware, IP or MAC address spoofing, DNS, dynamic host configuration protocol, and address resolution protocol. Attackers who get access to the system through MitM assaults can then manipulate and send data to other devices and users while posing as legitimate entities.

6.3.1   Eavesdropping Methods

Eavesdropping attacks typically involve using various tools to monitor network traffic and listen in on conversations. Pickup devices capture sounds or images from the attached microphones and cameras, which the attackers can then transform into an electrical format to listen in on their target. To eavesdrop, a communications link between a sender and a recipient is tapped.

Weak passwords make it simpler for attackers to access user accounts without authorization. Users who connect to open networks that don't require a password and don't encrypt data transmission provide hackers the perfect opportunity to listen in. Conversations overheard by telephone bugs are sent via a listening post. Due to their "always-on" status, which poses a serious risk to users' privacy, smart speech recognition assistants like Amazon Alexa and Google Home are also susceptible to listening; in 2011, the sniffing application Wireshark gave Android smartphone users many a headache.

Virtual meetings are as significant as in-person meetings. We increasingly rely on virtual meetings to fulfill our responsibilities as remote employment and the ethos of working from home become more prevalent. Attackers are using strategies like eavesdropping to intercept discussions and sensitive information discussed in virtual meetings, however, as they become more common. There are various impacts of an eavesdropping attack, such as:

Privacy loss: Every company and individual has personal information that, if made public, might damage their reputations.
Identity theft: Criminals can eavesdrop on login information they can use to get access to secure storage facilities. People don't just lose their identities.
Financial loss: Criminals with access to sensitive data can always threaten to make the information public unless the victim pays a significant quantity of money or sells it to rival businesses.

There are various techniques that can help prevent an eavesdropping attack.

Virtual private networks: The most popular type of eavesdropping protection, a virtual private network encrypts data between two points. The highest level of encryption should be used on corporate wireless networks.
Encryption: One of the greatest ways to guard against eavesdropping attempts is to encrypt data during transmission and during private discussions. Attackers are prevented from reading data sent between two parties through encryption. A good defense against eavesdropping assaults is military-grade encryption, which takes 500 billion years to decrypt.

Avoid suspicious links because they can be used by eavesdropping attackers to spread malware that includes eavesdropping malware. Maintain system updates because by taking advantage of flaws in outdated software, attackers can target users and organizations. It is crucial to upgrade your system to the most current security patches. Additionally, businesses should take physical security measures to safeguard the information in their offices from intruders who might physically bug desks, phones, and other gadgets. It is also necessary to create secure passwords and update them often, making sure they include a mix of capital and lowercase letters, numbers, and unusual characters; passwords should also be updated once a month to increase security. Additionally, if someone tries to eavesdrop on your discussion, a personal firewall will shield your data packets from them. Finally, only download software from reputable websites, such as the Apple or Google Play stores, as files acquired from other sources may contain malware that might install eavesdropping software without the user's knowledge.

Thanks to recent developments in chip fabrication techniques, computer makers can now include increasing CPU cores and other heterogeneous components on a System-on-Chip (SoC). Modern SoC designs are made up of a large number of IP cores that connect with one another via a Network-on-Chip (NoC) architecture. Because of its low latency and high throughput, NoC outperforms alternative communication methods regarding signal integrity, power usage, communication performance, and system scalability. As a result, manufacturers are now including NoCs more commonly in their SoC designs.

Manufacturers use the global supply chain to obtain third party IP cores from different regions of the world due to cost and time-to-market factors. The NoC has become a focus for potential security risks including the insertion of hardware Trojans (HTs) due to its dispersed design and growing usage. An implanted HT is capable of launching a wide range of assaults, including denial-of-service, eavesdropping, data integrity, spoofing, and side-channel assaults.

A standard threat model used to evaluate eavesdropping threats in NoCs is a router infected with HT working with a partner malicious program running on another IP core to intercept packets delivered via the router. Functional verification cannot reliably identify these HT-infected routers due to the NoC design's complexity. There are several ways to find these defective routers, including rule-based checkers included in wireless NoCs, additional validation checks, information obfuscation, and authentication and encryption. The main issue they frequently encounter is that they may add unacceptably high design overhead. Additionally, many of these methods rely on highly predictable traffic patterns and are unable to adapt to changing input or application requirements. When faced with realistic scenarios like changing application characteristics, job preemption, task migration, and input variances, they struggle.

6.3.2 ML Approaches

When other IP cores are connecting, an eavesdropping attacker listens in on sensitive information being transmitted over the NoC. These attackers seek to steal sensitive information by passively listening to on-chip communication (secure NoC packets), with the goal of leaking information over time without being noticed. We take into account a well-researched threat scenario where an ED assault is carried out by a hostile NoC router working with a second malicious software running on a rogue IP core. Fig. 1 gives a simple illustration of this using two trusted IP cores (B & H), and a malicious IP core (B), copies the packets received at the input buffer, alters the header information to direct the packets to E (an alternate destination), and then re-inserts the packets into the input buffer, makes it possible for this attack. We take into account cases where many pairs of sources and destinations are communicating at once while developing our threat model.

Several initiatives are being made to use ML principles to find HTs inside IP cores in the context of SoCs. The majority of ML techniques currently in use have been used to detect denial-of-service (DoS) assaults. For NoC-based many-core architectures, Kulkarni et al. made the initial attempt to identify real-time hardware Trojans [12]. Madden et al. [13] suggested a unique technique for identifying DoS assaults in NoCs that makes use of spiking neural networks; their detection method entailed evaluating the packet exchange between adjacent routers

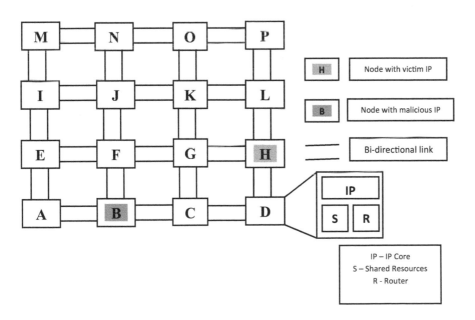

Figure 6.1 An ED Attack.

by looking at the traffic flow of a router's North, East, South, and West channels. Sudusinghe et al. [14] further developed the runtime detection of DoS attacks by training multiple ML models for various attack types and typical situations to recognize DoS attacks. To localize DoS attacks in NoC-based SoCs and shield accelerator-rich SoCs from attacks of this nature, Sinha et al. made a proposal [15]. Their strategy uses a localization technique and a ML approach to track the direction of the attack. To reduce performance and computational overheads, all of these methods used gathered data samples to offline train the corresponding ML models. It should be emphasized that the majority of the material currently in print focuses on applying ML to mitigate DoS attacks.

C. Sudusinghe et al. proposed a detection technique based on ML in N-o-C architectures [16]. The three phases of the approach are runtime, design time, and training time. The authors developed an eavesdropping-sensing algorithm (ESA) to detect ED attacks and designed router-attached probes to collect data from NoC packets with the least amount of performance and power overhead. Then, using the gem5 whole system simulator, they simulated both normal executions and attack executions. These operations collect NoC traffic data, which is then utilized to train ML models.

The decision unit (DU), a dedicated IP core, houses the taught ML models. The security policy engine in an NoC-based SoC shows how the SoC infrastructure may be used for computations related to a ML model. During runtime, the NoC traffic gathered by the probes is analyzed using the ML models accessible at the DU to draw conclusions about the router's data and certain time frames. The feature vectors within a particular time frame are pooled upon activation of the trigger mechanism and used as the input for the trained models. In this case, the ESA results will be used to determine which routers will be put up for evaluation. Then, using a collective decision-making method (CDMS), all of the selected routers are used to determine whether or not an ED attack is taking place. The result of the CDMS is the total probability of an attack occurring in the system. This combined ESA and CDMS technique reduces the possibility of false positives for typical network congestions because the ESA activates when the packet transfer ratio varies, and different routers are considered for CDMS.

In conclusion, because they may readily target phones, cellphones, and computers, eavesdropping assaults are a serious threat. Attackers may use this method to steal data for commercial gain, steal personal information to perpetrate identity theft, or use stolen login credentials to launch more serious attacks. Therefore, it is preferable to apply the preventive strategy indicated above now rather than regretting it later.

6.4   CHANNEL CONTAMINATION IN MASSIVE MIMO USING ML

For 5G and beyond, massive MIMO is one of the upcoming technologies [17, 18, 19, 20]. The fundamental idea behind the massive MIMO approach is that a base station (BS) can simultaneously service tens of user terminals (UTs) by being equipped with hundreds of antenna arrays. High spatial efficiency is achieved by spatial multiplexing to the UTs over the same time-frequency resource while using several antennas at the BS [17]. Massive MIMO systems offer many advantages, including high data speeds, great energy efficiency, and straightforward linear transceiver design [21].

To make the most of these advantages, the BS needs access to precise channel status information. Since the success of large MIMO systems depends on the accuracy of the estimation, channel estimation is a critical task. Because it is assumed that the channel response will be constant in both the time and frequency domains, the channel estimation is performed just once for each coherent block. By making use of the interchange between the uplink (UL) and downlink (DL) while employing a time division duplex (TDD), pilot signal broadcasts are restricted to the UL. The UL pilot signal transmissions, UL data signal transmissions, and DL data signal transmissions are the three groups into which the coherence block is separated.

Pilot signals should be as brief as feasible to increase spectrum efficiency but should increase as the number of UTs does to preserve the orthogonality to the signals of each UT. There are, therefore, a limited number of data transmission blocks available. To go around this issue, surrounding cells use the same pilot signal from the original cell. The BS cannot use copilots in the other cells to distinguish the signals from the UTs. Known also as inter-cell interference of the same pilot signals from the other cells, pilot contamination is the primary cause of channel estimation performance degradation [17].

Contrary to intra-cell interference, quick fading, and noise at the BS, the consequences of pilot contamination are difficult to counteract even if there are an infinite number of BS antenna arrays [21]. Many methods have been put forth to estimate channels in TDD-based large MIMO systems while pilot contamination is present. The minimum mean square error (MMSE) channel estimate is created using a covariance-aided channel estimation technique described in [22]. It is demonstrated that when the covariance matrices meet specific no-overlapping conditions, the pilot contamination is removed. The idea is used to offer a coordinated method for assigning the pilot signals, allowing the covariance matrices to be turned into the structure necessary to meet the need.

The BS lacks access to the channel covariance matrix, which expresses channel spatial correlation beforehand [23]. To estimate the covariance matrix, the strategy of adding extra pilots is suggested [24]; to some extent, it can take away the impacts of interference, because of the channel's time-varying nature, the amount of accessible coherence blocks for delivering data is constrained. Joint spatial division and multiplexing which lowers the overhead associated with channel estimation training, also requires this information [25]. There is no requirement for a priori knowledge in the blind method suggested in [26]. This method can tell the desired signal subspace apart from the interference subspace. It makes use of the sample covariance matrix's eigenvalue distribution, which distinguishes between intended and interfering signals after being derived from the received signals. As a result, it is feasible to arbitrarily distinguish between desirable and interfering signal subspaces. The power margin between the signals should be wide in order to discern the eigenvalue groups easily. Thus, when there is a sufficient power difference between the desired and interfering signals, pilot contamination can be reduced.

The authors of [27] suggested a methodology for a time-shifted pilot gearbox. Users of nearby cells receive DL data while a particular cell's users are broadcasting pilot signals. All users submit UL data once each group has finished the phase of sending pilot signals. Pilot contamination can be avoided with this technique. The users near the cell edge, however, experience interference from the BSs in the neighboring cells during the time of delivering pilot signals.

A pilot contamination precoding approach is suggested in [28]. Cooperation between several cells is used in this strategy. The inter-cell and intra-cell interference sum is minimized by the precoding matrix at the BS. Pilot contamination can be entirely eliminated when the number of BS antennas reaches infinity. However, this necessitates centralized processing at the BS, and each BS must be aware of all user data. The use of deep learning to tackle complex nonlinear problems in the fields of computer vision and natural language processing has grown in prominence in recent years including in the area of wireless communication [29]. The nonlinear optimization for channel estimation in large MIMO systems requires significant computing complexity. Because it can handle a range of non-convex and nonlinear issues, a deep learning-based technique is a contender for resolving this problem.

In [30], deep learning is integrated with direction-of-arrival estimation and channel estimation in massive MIMO systems. The plan is to collect sparsity characteristics in the angle domain and learn channel model statistics. The use of neural networks is said to help with channel estimation performance in [31]. Convolutional neural networks (CNNs) [32] are introduced as a class of low-complexity channel estimators, and this method is inspired by the structure of the MMSE channel estimator. In [33], it is demonstrated that CNN and long short-term memory can be used in the quick estimate of a time-varying channel.

The authors of [34] propose a deep learning-based two-stage channel estimation method that includes data- and pilot-aided stages. Signal identification and channel estimation follow an iterative structure in the data-aided stage; because data symbols are substantially longer than pilot symbols, the predicted channel quality is increased. Although pilot contamination is frequently present in multi-cell layout scenarios, all of these approaches [30, 31, 33, 34] assume a single-cell architecture and disregard its presence.

REFERENCES

[1] C. Szegedy, W. Zaremba, I. Sutskever, J. Bruna, D. Erhan, I. J. Goodfellow, and R. Fergus, "Intriguing properties of neural networks," in Proc. 2nd Int. Conf. Learn. Represent., Apr. 2014, pp. 1–10.

[2] I. J. Goodfellow, J. Shlens, and C. Szegedy, "Explaining and harnessing adversarial examples," in Proc. Int. Conf. Learn. Representations, Mar. 2015, pp. 1–11.

[3] B. Biggio, and F. Roli, "Wild patterns: Ten years after the rise of adversarial machine learning," Pattern Recognit., vol. 84, pp. 317–331, Dec. 2018.

[4] N. N. Dalvi, P. M. Domingos, Mausam, S. K. Sanghai, and D. Verma, "Adversarial classi_cation," in Proc. 10th ACM SIGKDD Int. Conf. Knowl. Discov. Data Min., Aug. 2004, pp. 99–108.

[5] B. Nelson, M. Barreno, F. J. Chi, A. D. Joseph, B. I. P. Rubinstein, U. Saini, C. A. Sutton, J. D. Tygar, and K. Xia, "Exploiting machine learning to subvert your spam _lter," in Proc. USENIX Workshop Large-Scale Exploit. Emerg. Threat., Apr. 2008, pp. 1–9.

[6] M. Barreno, B. Nelson, A. D. Joseph, and J. D. Tygar, "The security of machine learning," Mach. Learn., vol. 81, no. 2, pp. 121–148, Nov. 2010.

[7] N. Akhtar, and A. S. Mian, "Threat of adversarial attacks on deep learning in computer vision: A survey," IEEE Access, vol. 6, pp. 14410–14430, Jul. 2018.

[8] X. Yuan, P. He, Q. Zhu, and X. Li, "Adversarial examples: Attacks and defenses for deep learning," IEEE Trans. Neural Netw. Learn. Syst., vol. 30, no. 9, pp. 2805–2824, Sep. 2019.

[9] M. S. Riazi, and F. Koushanfar, "Privacy-preserving deep learning and inference," in Proc. Int. Conf. Comput.-Aided Design ICCAD, Nov. 2018, pp. 1–4.

[10] Q. Liu, P. Li, W. Zhao, W. Cai, S. Yu, and V. C. M. Leung, "A survey on security threats and defensive techniques of machine learning: A data driven view," IEEE Access, vol. 6, pp. 12103–12117, 2018.

[11] N. Papernot, P. D. McDaniel, A. Sinha, and M. P. Wellman, "SoK: Security and privacy in machine learning," in Proc. IEEE Eur. Symp. Secur. Privacy (EuroS&P), Apr. 2018, pp. 399–414.

[12] A. Kulkarni et al., "Svm-based real-time hardware trojan detection for many-core platform," in ISQED, 2014.

[13] K. Madden et al., "Adding security to networks-on-chip using neural networks," in SSCI, 2018.

[14] C. Sudusinghe et al., "Denial-of-service attack detection using machine learning in network-on-chip architectures," in International Symposium on Networks-on-Chip (NOCS), 2021, pp. 35_40.

[15] M. Sinha et al., "Sniffer: A machine learning approach for dos attack localization in noc-based socs," JETCAS, pp. 278–291, 2021.

[16] C. Sudusinghe, S. Charles, S. Ahangama, and P. Mishra, "Eavesdropping attack detection using machine learning in network-on-chip architectures," IEEE Design & Test, vol. 39, no. 6, pp. 28–38, Dec. 2022, doi: 10.1109/MDAT.2022.3202995.

[17] T. L. Marzetta, "Noncooperative cellular wireless with unlimited numbers of base station antennas," IEEE Trans. Wireless Commun., vol. 9, no. 11, pp. 3590–3600, Nov. 2010.

[18] E. G. Larsson, O. Edfors, F. Tufvesson, and T. L. Marzetta, "Massive MIMO for next generation wireless systems," IEEE Commun. Mag., vol. 52, no. 2, pp. 186–195, Feb. 2014.

[19] L. Lu, G. Y. Li, A. L. Swindlehurst, A. Ashikhmin, and R. Zhang, "An overview of massive MIMO: Benefits and challenges," IEEE J. Sel. Top. Signal Process., vol. 8, no. 5, pp. 742–758, Oct. 2014.

[20] E. Björnson, E. G. Larsson, and M. Debbah, " Massive MIMO for maximal spectral efficiency: How many users and pilots should be allocated?" IEEE Trans. Wireless Commun., vol. 15, no. 2, pp. 1293–1308, Feb. 2016.

[21] O. Elijah, C. Y. Leow, T. A. Rahman, S. Nunoo, and S. Z. Iliya, "A comprehensive survey of pilot contamination in massive MIMO—5G system," IEEE Commun. Surv. Tuts., vol. 18, no. 2, pp. 905–923, Apr.– Jun. 2016.

[22] H. Yin, D. Gesbert, M. Filippou, and Y. Liu, "A coordinated approach to channel estimation in large-scale multiple-antenna systems," IEEE J. Sel. Areas Commun., vol. 31, no. 2, pp. 264–273, Feb. 2013.

[23] S. Haghighatshoar, and G. Caire, "Massive MIMO pilot decontamination and channel interpolation via wideband sparse channel estimation," IEEE Trans. Wireless Commun., vol. 16, no. 12, pp. 8316–8332, Dec. 2017.

[24] E. Björnson, L. Sanguinetti, and M. Debbah, "Massive MIMO with imperfect channel covariance information," in Proc. 50th Asilomar Conf. Signals, Syst., Comput., Nov. 2016, pp. 974–978.

[25] A. Adhikary, J. Nam, J. Ahn, and G. Caire, "Joint spatial division and multiplexing— The large-scale array regime," IEEE Trans. Inf. Theory, vol. 59, no. 10, pp. 6441–6463, Oct. 2013.

[26] R. R. Müller, L. Cottatellucci, and M. Vehkaperä, "Blind pilot de-contamination," IEEE J. Sel. Top. Signal Process., vol. 8, no. 5, pp. 773–786, Oct. 2014.

[27] F. Fernandes, A. Ashikhmin, and T. L. Marzetta, "Inter-cell interference in noncooperative TDD large scale antenna systems," IEEE J. Sel. Areas Commun., vol. 31, no. 2, pp. 192–201, Feb. 2013.

[28] J. Jose, A. Ashikhmin, T. L. Marzetta, and S. Vishwanath, "Pilot contamination and precoding in multi-cell TDD systems," IEEE Trans. Wirel. Commun., vol. 10, no. 8, pp. 2640–2651, Aug. 2011.

[29] H. Huang et al., "Deep learning for physical-layer 5G wireless techniques: Opportunities, challenges and solutions," IEEE Wireless Commun., vol. 27, no. 1, pp. 214–222, Feb. 2020.

[30] H. Huang, J. Yang, H. Huang, Y. Song, and G. Gui, "Deep learning for super-resolution channel estimation and DOA estimation based massive MIMO system," IEEE Trans. Veh. Technol., vol. 67, no. 9, pp. 8549–8560, Sep. 2018.

[31] D. Neumann, T. Wiese, and W. Utschick, "Learning the MMSE channel estimator," IEEE Trans. Signal Process., vol. 66, no. 11, pp. 2905–2917, Jun. 2018.

[32] Y. LeCun, P. Haffner, L. Bottou, and Y. Bengio, Object Recognition With Gradient-Based Learning. Berlin, Germany: Springer-Verlag, 1999.

[33] Y. Liao, Y. Hua, X. Dai, H. Yao, and X. Yang, "ChanEstNet: A deep learning based channel estimation for high-speed scenarios," in Proc. IEEE Int. Conf. Commun., May 2019, pp. 1–6.

[34] C. Chun, J. Kang, and I. Kim, "Deep learning-based channel estimation for massive MIMO systems," IEEE Wireless Commun. Lett., vol. 8, no. 4, pp. 1228–1231, Aug. 2019.

Chapter 7

Traffic Prediction and Congestion Control Using Regression Models in Machine Learning for Cellular Technology

R. Madonna Arieth, Subrata Chowdhury,
B. Sundaravadivazhagan, and Gautam Srivastava

7.1 INTRODUCTION

Because of increasing populations worldwide, there has been a need for rapid changes in mobile communication technology, and there has lately been explosive development of cellular technologies [1]. Mobile device users have demanded higher data rates and quality of service, prompting network service providers to seek new technologies. The fourth-generation (4G) mobile network improved over the earlier three generations [4], but countries have now been using the fifth generation of the network (5G), although this technology is not yet widespread worldwide. The 5G network is designed to adapt a wide range of services with increased mobile broadband, larger machine-type communications, and ultra-reliable low-latency communications, as well as to meet the bandwidth requirements of subscribers [5]. Here are several key features and aspects of 5G:

- Speed and capacity: Compared with earlier generations, 5G offers higher data transfer rates. It can provide blossom speeds of up to 10 gigabits per second (Gbps), This high speed allows for seamless streaming of high-definition videos, faster file downloads, and smoother online gaming experiences. Additionally, 5G provides more network capacity, allowing more devices to connect simultaneously.
- Low latency: Latency is the duration of time it takes for data to transfer between devices on a network, and 5G offers ultra-lower latency, reducing delays to as short as 1 millisecond (ms). This near-real-time responsiveness is crucial for applications that require immediate interaction.
- Increased connectivity: The goal of 5G is to connect many devices and objects simultaneously and continuously.
- Network slicing: With 5G, network operator can build numerous virtual networks on a single physical infrastructure. Each network slice can be tailored to specific applications or services, providing customized performance characteristics,

DOI: 10.1201/9781003306290-7

89

security, and quality of service. This flexibility enables diverse use cases with different requirements to coexist on the same network infrastructure.

- Enhanced coverage: 5G is designed to provide broader coverage, including in both densely populated urban areas and remote rural regions. It utilizes low-, mid-, and high-frequency bands, to achieve a balance between coverage and capacity. Lower-frequency bands enable wider coverage, while higher-frequency bands offer faster speeds and more capacity but with shorter range.
- Enabling new technologies: 5G is expected to be a catalyst for numerous technological advancements like self-driving vehicles, smart cities, modern healthcare systems, virtual reality, augmented reality, and artificial intelligence (AI). These developments could revolutionize industries, enhance quality of life, and spur economic expansion [6].

It's important to remember that because 5G requires significant infrastructure upgrades, the deployment and availability of networks and services may differ across different regions and countries. However, as 5G continues to roll out globally, it holds the promise of revolutionizing the way we use technology for communication, work, and other activities in a new era of connectivity and innovation.

One of the important challenges in 5G is data communication [7]. With 4G, mobile traffic congestion has been a problem [2, 3, 8], and the 5G network claims to have solved the issue. To improve a network's quality of service (QoS), precise time-series models that forecast cellular mobile traffic have become crucial [12, 15, 21, 26, 28 & 30]. Regression models based on machine learning algorithms have been highly developed [9, 11, 13, 22, 27] and utilized for mobile data traffic prediction in order to increase user experience and better utilize resources (see Figure 7.1).

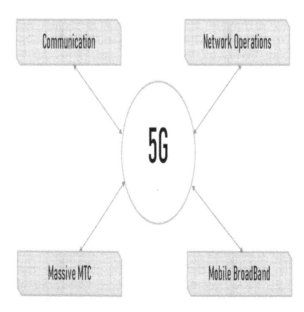

Figure 7.1 Overview of the 5G Network.

7.2   DATA TRANSFER IN CELLULAR 5G

The 5G network transmits data through various technologies and protocols that transfer the data with high speed and reliable communication. The process begins with radio access networks (RANs), which consist of base stations called gNodeBs (gNBs) that are accountable for transmitting and receiving wireless data to and from user devices; after that interface is the wireless communication link between the gNB and user devices. The 5G technology is used to optimize spectral efficiency and transfer the data to the core network once it data is received at the gNB.

The midpoint of network serves the backbone of the 5G network and handles various tasks, including routing, authentication, and traffic management. The network is sliced, which means dividing the core network into virtual networks. Each slice is made to order of specific applications or services, providing dedicated resources, quality of service, and security parameters. Data from different slices is kept separated and maintained separately within the core network. Beyond that, 5G networks are based on an IP (Internet protocol) architecture, allowing broadloom integration with existent Internet infrastructure. It enables streamlined data transfer between different network components and alleviate connectivity with other IP-based systems and services, and then 5G networks employ a breakup of user plane and control plane architecture. The control plane handles tasks related to network management, signaling, and control, while the user plane is responsible for data forwarding. This separation allows for more flexibility, scalability, and efficient resource allocation.

In contrast to processing data in centralized data centers, edge computing processes data closer to the network edge. This helps reduce latency and offloads processing tasks from the core network. The 5G networks employ the virtualization of network functions, such as routing, firewall, and load balancing. They can dynamically allocate resources, scale services, optimize network operations, and maintain the QoS. The 5G networks prioritize different types of data traffic based on quality of service requirements. QoS management ensures that time-sensitive and critical applications, such as autonomous vehicles or remote surgeries, receive higher priority and better network resources to guarantee low latency and high reliability. Overall, 5G networks employ advanced wireless technologies, IP-based architecture, network slicing, edge computing, and virtualization to enable high-speed, low-latency, and efficient data transfer. These features contribute to the enhanced user experience and support a broad extent of applications and services

7.3   TRAFFIC IN 5G

In 5G networks, traffic refers to the data and communication exchanged between devices, users, and applications over the network. Multiple factors influence the increased speed, capacity, and connectivity capabilities of 5G:

> Increased data consumption: 5G's faster speeds and enhanced user experience encourage higher data usage. Users can stream high-definition videos, download large files, and engage in data-intensive applications more

seamlessly. This increased data consumption contributes to higher traffic volumes on the network.

Internet of Things (IoT) devices: 5G enables a massive number of IoT devices to connect simultaneously. These devices, such as sensors, smart appliances, and industrial machines, generate a significant amount of data traffic as they communicate with each other and transmit information to the cloud or other devices.

Mobile broadband and multimedia services: 5G offers enhanced mobile broadband, allowing users to enjoy high-quality multimedia services on their mobile devices. This includes streaming video content, video conferencing, online gaming, and virtual reality applications. These bandwidth-intensive services contribute to increased traffic demands on the network.

Cloud services and edge computing: 5G networks enable faster access to cloud services and support edge computing capabilities. Cloud-based applications and services, such as file storage, software-as-a-service (SaaS), and real-time collaboration tools, generate traffic as data is transmitted to and from the cloud. Similarly, edge computing offloads processing tasks closer to the network edge, reducing latency and network congestion while generating additional traffic within the network.

Network slicing and service differentiation: Network slicing enables network operators to build numerous virtual networks with different characteristics to cater to specific services and applications. Each network slice can have different bandwidth, latency, and QoS parameters. This service differentiation and customization contribute to diverse traffic patterns within the 5G network.

Application-specific traffic: Different applications have varying traffic requirements and characteristics. For example, video streaming applications generate continuous and high-bandwidth traffic, while messaging applications involve smaller intermittent data exchanges. The mix of applications used by users and the corresponding traffic generated by each application contribute to the overall traffic pattern in a 5G network

Time and location: Traffic patterns are different query based on period of time and location. In the time of peak when users are streaming videos or accessing social media, network traffic tends to be higher. Additionally, in densely populated areas, such as urban centers or event venues, network traffic can experience spikes due to high concentrations of users.

It's worth noting that the specific traffic characteristics and patterns in a 5G network can vary based on the deployment, user behavior, and the applications and services being used. Network operators continually monitor and optimize their networks to handle the evolving traffic demands and ensure efficient network performance. While data transfer in 5G networks is designed to be efficient and reliable, there are certain challenges and problems that can occur. One of the main issues with data transfer in 5G is congestion.

7.4 CONGESTION IN 5G

In 5G networks, congestion can occur during data transfer, leading to degraded performance and slower data speeds. Here are some common types of congestion that can occur in 5G:

RAN congestion: RAN congestion happens when the level of data flow or the number of connected devices exceeds the capacity of the base stations (gNodeBs). This can occur in densely populated areas or during peak usage periods, such as in crowded stadiums or during major events. RAN congestion can result in slower data speeds, increased latency, and even dropped connections.

Backhaul congestion: Backhaul refers to the network connections that link the gNodeBs to the core network, and congestion occurs when the backhaul capacity is insufficient to handle the volume of data traffic from the gNodeBs. Inadequate backhaul capacity can lead to bottlenecks and slower data transfer rates, impacting overall network performance.

Core network congestion: Core network congestion occurs when the core network infrastructure, including switches, routers, and servers, is overwhelmed with data traffic. This can happen due to a sudden surge in user demand or when multiple services and applications require high network resources simultaneously. Core network congestion can result in increased latency, packet loss, and reduced data transfer speeds.

Interference congestion: Interference congestion arises when there is interference in the radio spectrum used by 5G networks. This interference can be caused by other wireless devices, neighboring networks [25], or environmental factors. Interference congestion can lead to decreased signal quality, weaker coverage, and degraded data transfer performance.

Application-specific congestion: Certain applications or services that generate high-volume data traffic can cause congestion within the network. For example, video streaming, large file downloads, or cloud-based services can consume significant bandwidth and put strain on network resources. When multiple users simultaneously engage in bandwidth-intensive applications, it can result in localized congestion and slower data transfer speeds.

Virtual network slice congestion: In 5G, slicing congestion allows the virtual networks to dedicate resources for specific applications or services. If the resources allocated to a particular network slice are insufficient to handle the traffic demands, congestion can occur within that slice. This can affect the performance of applications relying on that slice and impact the data transfer experience.

Network congestion: High network usage or spikes in data traffic can lead to network congestion, causing lower speed and longer latency. This can occur in densely populated areas or during peak usage hours. Network operators continuously monitor and manage their networks to mitigate

congestion issues employing techniques such as traffic shaping, load balancing, dynamic resource allocation, and capacity planning.

Signal interference and coverage limitations: The higher-frequency bands
used in 5G networks have shorter ranges than the lower-frequency bands.
Obstacles such as buildings, trees, and geographical terrain can cause signal
interference and impact the coverage and quality of data transfer. Network
operators deploy a combination of small cells, repeaters, and other infrastructure enhancements to address coverage limitations and ensure reliable
data transfer.

Reliability and availability: Despite the advancements in 5G technology, occa-
sional service disruptions or outages can still occur due to equipment failures, power outages, natural disasters, or maintenance activities.

Security and privacy: Data security and privacy are critical concerns in data
transfer. The 5G networks implement strapping security such as encryption and authentication, and network operators, service providers, and users
must adopt best practices in security and privacy to mitigate these risks.

Compatibility and interoperability: 5G networks need to interoperate with various devices, technologies, and legacy networks. Compatibility issues can
arise when connecting older devices or when transitioning from previous
network generations. Standardization efforts and compatibility testing help
address these challenges, ensuring seamless data transfer across different
devices and networks.

QoS variations: Different applications and services have varying requirements
for bandwidth, latency, and reliability. Ensuring consistent QoS for all
applications can be challenging. Network operators employ mechanisms
like traffic prioritization, QoS management, and network slicing to allocate
resources effectively and meet diverse application requirements.

Data caps and fair usage policies: Some 5G service plans impose data caps or
fair usage policies that limit the amount of data users can transfer within
a specific time period. Exceeding these limits can result in reduced data
transfer speeds or additional charges. Users should be aware of their plan's
limitations to avoid disruptions in data transfer.

It's important to note that network operators continuously work to address
these challenges and improve the entire performances and reliability of 5G networks. As technology evolves and network infrastructure expands, many of these
issues are expected to be mitigated or resolved to provide a better data transfer
experience for user.

7.5 TRAFFIC CONGESTION IN 5G

In 5G networks, traffic congestion occurs when the data transfer from source to
destination exceeds the capacity of network, leading to reduced performance,

dragging data speeds, and extended latency. Here are some factors that can contribute to traffic congestion in 5G [17]:

Increased data demand: 5G offers significantly faster speeds and enhanced user experiences, leading to a surge in data consumption. With the growing of high-bandwidth applications, the overall demand for data increases. This surge in data demand can put strain on the network infrastructure and potentially lead to congestion.

Growing number of connected devices: 5G supports large numbers of integrated devices at the same time. As more sensors, wearable, and smart home devices join the network, overall data traffic increases. The influx of data generated by these devices can contribute to congestion if the network capacity is not scaled accordingly.

Dense urban areas: Urban areas with high population densities tend to experience more network congestion due to the concentrations of users and their data demands. In city centers, transportation hubs, or crowded public spaces, the simultaneous usage of data-intensive applications by a large number of users can strain the network capacity and result in congestion.

Peak usage periods: During peak usage periods, such as evenings or weekends when a significant number of users engage in bandwidth-intensive applications simultaneously, the network can experience congestion. The increased demand during these periods can exceed the available network capacity and lead to congestion.

Insufficient infrastructure: In some cases, the network infrastructure may not be adequately designed or deployed to handle the anticipated traffic demands. Insufficient base stations, backhaul connections, or core network resources can limit the network's capacity and contribute to congestion. Network operators need to ensure proper infrastructure planning and capacity upgrades to accommodate growing traffic demands.

Spectrum limitations: The availability of sufficient radio spectrum is crucial for supporting high-capacity data transfer in 5G networks. If the available spectrum is limited or congested, it can affect the network's ability to handle increased data traffic. Spectrum management, efficient spectrum allocation, and spectrum sharing mechanisms are essential for mitigating congestion related to spectrum limitations.

To address traffic congestion in 5G networks, operators employ various strategies such as optimizing network resource allocation, implementing traffic management techniques, deploying additional infrastructure, and upgrading network capacity. Additionally, advanced technologies like network slicing, QoS management, and dynamic resource allocation help prioritize critical applications and ensure efficient utilization of network resources, reducing the impact of congestion on data transfer. We are proposing regression in machine learning techniques [16] to overcome mobile data traffic congestion.

7.6   REGRESSION MODELS IN MACHINE LEARNING

Regression models based on machine learning algorithms have been formulated for the prediction of mobile traffic [27] to improve resource usage and amend the user experience (Figure 7.2). These models aim to forecast the future traffic patterns and demands, allowing network operators to adapt their infrastructure and allocate resources efficiently [18] and contain a number of features that contribute to the task.

Data collection: Mobile data traffic prediction models rely on historical data collected from the network, including user activity, data usage patterns, time of day, location, and network performance metrics. This data serves as the training set for the regression models.

Feature selection: A very important step in building regression models for mobile data traffic prediction. Relevant features from the historical data are identified and given as input variables in the regression models to generate accurate predictions.

Training and validation: The regression models are trained using historical data, with a portion of the data set aside for validation and testing. The models are optimized and adjusted to minimize prediction errors and improve accuracy and evaluated via cross-validation.

Traffic forecasting: Once the regression models are trained and validated, they can be used to forecast mobile data traffic. The models take into account various factors such as time of day, location, and historical trends to predict future traffic volumes and patterns. This information assists network operators in optimizing resource allocation, capacity planning, and network management to meet the anticipated traffic demands

Real-time adaptation: Mobile data traffic prediction models can be continuously updated and refined based on real-time data feedback. As new data becomes available, the models can be adjusted to reflect changing traffic patterns and ensure accurate predictions.

By employing regression models in mobile data traffic prediction, network operators can make informed decisions to optimize their resources, mitigate congestion, improve network performance, and provide a better user experience. These models assist in efficient network planning, capacity optimization, and proactive management of the network infrastructure. There are different congestion prediction methods, including decision trees, logistic regression, and neural networks, with different performance features.

Decision trees are machine learning models [18, 19, 20] that predict by recursively dividing the data based on feature conditions. Decision trees split the information in accordance with the values of various attributes to produce a decision-tree-like model. They use information gain or the Gini index to determine the best attribute for splitting. They are intuitive and easygoing to interpret, providing a transparent

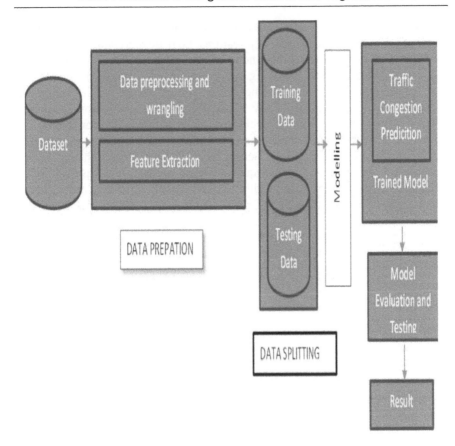

Figure 7.2 Traffic Model.

decision-making process. In the context of traffic prediction, decision trees can identify important features and patterns to estimate traffic levels. However, decision trees may struggle with capturing complex relationships and tend to overfit the training data, leading to reduced generalization. Decision trees aim to create a model that predicts traffic congestion based on relevant features and attributes.

The main advantage of decision trees is that they are easy to understand and interpret. They can capture nonlinear relationships and interactions between features. Decision trees can identify the most important features for predicting traffic congestion, visualize the flow of predictions, and provide decision rules. Performance measures for decision trees include accuracy, precision, recall, F1 score, and area under the receiver operating characteristic curve.

Another data traffic congestion prediction method is logistic regression. By simulating the link between input data and a binary response variable, logistic regression can be utilized to predict binary outcomes, approximating the probability of congestion based on relevant features. Logistic regression estimates the coefficients that maximize the likelihood of the observed data. It entails interpretation and is suitable for analyzing the impact of individual predictors. However, logistic regression assumes a linear relationship between predictors and the response, which may limit its ability to capture complex congestion dynamics. Its advantages are that it is interpretable and provides probabilistic predictions, it can handle both categorical and continuous variables, and it performs well with small to moderate-sized data sets. Logistic regression can identify the significant predictors of traffic congestion and quantify their impacts. It can provide odds ratios for interpreting the effects of different variables. Performance measures for logistic regression include accuracy, precision, recall, F1 score, area under the ROC curve, and log-loss.

Neural networks are another method of predicting data traffic congestion; they are a subset of artificial intelligence that is capable of learning complex patterns and nonlinear relationships from large datasets. Neural networks consist of interconnected layers of artificial neurons that will process the input data. They excel at capturing intricate traffic patterns and congestion dynamics by automatically extracting features from the data. However, neural networks often require substantial amounts of training data and computational resources for training, and their complex structure makes them less interpretable compared to simpler models like decision trees or logistic regression.

The major objective of neural networks in this context is traffic prediction and congestion control by learning complex patterns and relationships in the data [10]. The underlying methodology consists of coordinated layers of neurons that process and propagate information. The networks use activation functions and back propagation algorithm to learn the weights and biases. The advantages of neural networks are that they can capture nonlinear and intricate relationships in the data, handle large amounts of data and high-dimensional features, and adapt to changing patterns over time. Neural networks can make accurate predictions of traffic congestion based on historical data; they can learn temporal dependencies and identify patterns that are difficult for other models to capture. Performance measures for neural networks depend on the specific architecture and task. Common measures include accuracy, mean squared error, root mean squared error, and mean absolute error.

7.7 COMPARATIVE STUDY CONSIDERATIONS

When comparing these prediction methods for traffic and congestion, several factors should be considered:

Performance: Evaluate the accuracy and predictive capability of each method by comparing their performance metrics. Assess how well each method predicts traffic levels or congestion states.

Model complexity: Consider the complexity and interpretability of each method. Decision trees and logistic regression models are relatively simple and offer interpretability, making them suitable for gaining insights into feature importance. Neural networks, while powerful, are more complex and less interpretable.

Scalability: Assess the scalability of the methods, especially when dealing with large-scale traffic or congestion data sets. Neural networks require more computational resources and longer training times than decision trees or logistic regression.

Data requirements: Consider the data requirements for each method. Decision trees and logistic regression can work well with smaller data sets, To operate at their best, neural networks frequently need bigger volumes of labeled training data.

Adaptability: Assess how well each method can adapt to changing traffic or congestion patterns over time. Some methods require retraining or adjustment of parameters to maintain accuracy as traffic conditions evolve.

Ultimately, the choice of data congestion prediction method depends on the specific requirements of the application, the available data, and the desired trade-offs between model complexity, interpretability, and predictive performance. It is often beneficial to experiment with multiple methods and evaluate their performance on relevant datasets to determine the most suitable approach for traffic and congestion prediction.

7.8 CONCLUSION

Three models, logistic regression, decision tree, and neural network, are the current tools for predicting and controlling data traffic in mobile networks. Their features such as accuracy, interpretability, ability to handle different types of data, and capacity to capture complex relationships can be compared to determine which will be most effective for the task; the specific choice of model will depend on the characteristics of the traffic data, the level of interpretability required, and the trade-off between prediction accuracy and computational complexity logistic regression might be suitable. If there are complex interactions and interpretability is important, decision trees or random forests can be considered. If there are complex patterns and larger amounts of data available, neural networks and deep learning models, may provide better prediction performance with lower interpretability.

REFERENCES

1. H. Haile et al., "End-to-end congestion control approaches for high throughput and low delay in 4G/5G cellular networks," Computer Networks, vol. 186, 26 February 2021, p. 107692.

2. J. Lorincz et at., "A comprehensive overview of TCP congestion control in 5G networks: Research challenges and future perspectives," Sensors, vol. 21, no. 13, p. 4510, 2021, https://doi.org/10.3390/s21134510

3. S. Mendonç et al., "The rise of 5G technologies and systems: A quantitative analysis of knowledge production," Telecommunications Policy, vol. 46, no. 4, p. 102327, May 2022.

4. M. Attaran etal., "The impact of 5G on the evolution of intelligent automation and industry digitization," Journal of Ambient Intelligence and Humanized Computing, vol. 14, pp. 5977–5993, 2023.

5. M. A. Albreem, (2015, April). 5G wireless communication systems: Vision and challenges. In 2015 International Conference on Computer, Communications, and Control Technology (I4CT) (pp. 493–497). IEEE.

6. S., Mane, "5G communications & networks," International Journal of All Research Education and Scientific Methods (IJARESM), vol. 10, no. 9, ISSN: 2455–6211, September 2022.

7. N. O. Nwazor, and V. K. Ugah, "Device-To-Device (D2D) data communications in 5G networks," International Journal of Advances in Engineering and Management (IJAEM), vol. 4, no. 1, pp. 1151–1154, January 2022.

8. A. H. Tanveer, R. Ali, and A. Kim, "Machine learning for physical layer in 5G and beyond wireless networks: A survey," Electronics, vol. 11, p. 121, 2022. https://doi.org/10.3390/electronics11010121

9. S. Sarkar, "Aritri debnath, machine learning for 5G and beyond: Applications and future directions, IEEE Electronics and Sustainable Communication Systems (ICESC)," Second International Conference, 4–6 August 2021. DOI: 10.1109/ICESC51422.2021

10. S. E. A. Alnawayseh et al., Smart Congestion Control in 5G/6G Networks Using Hybrid Deep Learning Techniques, Hindawi Volume, 2022| Article ID 1781952 | https://doi.org/10.1155/2022/1781952

11. D. Alekseeva, N. Stepanov, A. Veprev et al., "Comparison of machine learning techniques applied to traffic prediction of real wireless network," IEEE Access, vol. 9, pp. 159495–159514, 2021.

12. C. Zhang, H. Zhang, D. Yuan, and M. Zhang, "Citywide cellular traffiffiffic prediction based on densely connected convolutional neural networks," IEEE Communications Letters, vol. 22, no. 8, pp. 1656–1659, 2018.

13. R. M. Yasir, Dr. N. Nower, Dr. M. Shoyaib, Traffic Congestion Prediction Using Machine Learning Techniques, arXiv: 2206.10983v3 [cs.LG], 7 September 2022.

14. Y. J. Wu, and M. Xu, "Traffic flow prediction with rainfall impact using a deep learning method," Journal of Advanced Transportation, 2017.

15. J. Feng, X. Chen, R. Gao, M. Zeng, and Y. Li, "Deep TP: An end-to-end neural network for mobile cellular traffiffiffic prediction," IEEE Network, vol. 32, no. 6, pp. 108–115, 2018.

16. C. L. Chowdhary, & P. C. Mouli, (2012, March). Design and implementation of secure, platform-free, and network-based remote controlling and monitoring system. In International Conference on Pattern Recognition, Informatics and Medical Engineering (PRIME-2012) (pp. 195–198). IEEE.

17. H. D. Trinh, L. Giupponi, and P. Dini, "Mobile traffiffiffic prediction from raw data using LSTM networks," in Proceedings of the IEEE 29th Annual International Symposium on Personal, Indoor and Mobile Radio Communications (PIMRC), pp. 1827–1832, Bologna, Italy, September 2018.

18. J. Zhao, H. Qu, J. Zhao, & D. Jiang, (2018). Towards traffic matrix prediction with LSTM recurrent neural networks. Electronics Letters, 54(9), 566–568.
19. A. S. Khatouni, F. Soro, and D. Giordano (editors), "A machine learning application for latency prediction in operational 4g networks," IFIP/IEEE Symposium on Integrated Network and Service Management (IM), pp. 71–74, Arlington, VA, USA: IEEE, 8–12 April 2019.
20. A. Patel, Machine Learning Algorithm Overview [Internet] [cited 2020 May 9], 2018. Available from: https://medium.com/mlresearch-lab/machine-learning-algorithm-overview-5816a2e6303
21. R. Li, Z. Zhao, J. Zheng et al., "The learning and prediction of application-level traffic data in cellular networks," IEEE Transactions on Wireless Communications, vol. 16, no. 6, pp. 3899–3912, 2017.
22. A. E. Mohamed, "Comparative study of four supervised machine learning techniques for classification," International Journal of Applied, vol. 7, no. 2.
23. www.uctel.co.uk/blog/whats-the-difference-between-3g-4g-and-5g
24. R. Dong, C. She, W. Hardjawana, Y. Li, and B. Vucetic, "Deep learning for hybrid 5G services in mobile edge computing systems: Learn from a digital twin," IEEE Transactions on Wireless Communications, vol. 18, no. 10, pp. 4692–4707, 2019.
25. S. A. R. Zaidi, "Nearest neighbour methods and their applications in the design of 5G and beyond wireless networks," ICT Express, vol. 7, no. 4, pp. 414–420, 2021.
26. M. Cosovic, A. Tsitsimelis, D. Vukobratovic, J. Matamoros, and C. Anton-Haro, "5G mobile cellular networks: Enabling distributed state estimation for smart grids," IEEE Communications Magazine, vol. 55, no. 10, pp. 62–69, 2017.
27. I. A. Najm, A. K. Hamoud, J. Lloret, and I. Bosch, "Machine learning prediction approach to enhance congestion control in 5G IoT environment," Electronics, vol. 8, no. 6, p. 607, 2019.
28. X. C. Feng, R. Gao, M. Zeng, and Y. Li, "DeepTP: An end-to-end neural network for mobile cellular traffic prediction," IEEE Network, vol. 32, no. 6, pp. 108–115, 2018.
29. L. Nie, D. Jiang, S. Yu, and H. Song, "Network traffic prediction based on deep belief network in wireless mesh backbone networks," in Proceedings of the 2017 IEEE Wireless Communications and Networking Conference (WCNC), pp. 1–5, San Francisco, CA, USA, March 2017.
30. K. Udit, and S. Debarshi, "An overview of device-to-device communication in cellular networks," ScienceDirect—ICT Express, vol. 4, pp. 203–208, 2018.

Chapter 8

Resource Allocation Optimization

Monika K, Chandrakumar T, B. Sundaravadivazhagan, and Ramya Govindaraj

8.1 INTRODUCTION

From the first generation of mobile communication technology, which appeared in the 1980s, to the fifth generation of mobile communication technology, wireless technology has evolved from satellite-based communications to radio broadcasting, and then eventually developed into intelligent terminal devices. As a result, wireless communication technology is now fully capable of providing general voice communication or simple data services as well as being fully integrated into other types of technology [1]. It has evolved into an essential component of contemporary society, improving the convenience and quality of people's lives. Fourth-generation (4G) wireless technology's launch offered a foundation for the transition to higher data speeds and dependable communication standards [2].

The long-term evolution (LTE) of wireless communication standards and the emergence of global interoperability for microwave access are both results of the rising demand for data services. However, it is extremely challenging to handle the quickly rising data rates and connected devices in the existing 4G LTE cellular infrastructure due to the rise in smart terminal devices, the exponential rise in wireless data consumption and usage, and the introduction of developing multimedia applications. While the wireless network scales the system capacity primarily by expanding the number and density of base station deployments, in traditional radio access network architecture, the base station controller performs data transmission and reception by controlling the RF units; consequently, the expansion of the number and density of base stations poses many challenges and difficulties [3].

In 2024, the progression of 5G technology in resource allocation optimization has seen significant strides. With continued advancements in artificial intelligence and machine learning algorithms, telecom companies and researchers have been able to refine and enhance the efficiency of resource allocation in 5G networks. Dynamic spectrum sharing techniques have become more sophisticated, allowing for better utilization of available frequency bands. Moreover, intelligent network management systems leveraging predictive analytics have emerged, enabling proactive resource allocation adjustments based on real-time traffic demands and network conditions. These developments have resulted in improved network

DOI: 10.1201/9781003306290-8

performance, higher throughput, reduced latency, and enhanced overall user experience in 5G environments [4, 5].

The Cisco Visual Index predicts that by 2021, the average amount of Internet traffic per person will be 30 GB, with more than 63% of that traffic coming from wireless and mobile devices. The global trends of cloud computing and the Internet of Things (**IoT**), which digitizes our world, are particularly to blame for this increase. As robotics and virtual transformation become more prevalent in our lives, intelligent and immersive experience maintenance will be necessary [6]. Massive multiple-input/multiple-output (MIMO) beamforming is a technology used in mobile communication systems to enhance the performance of wireless networks. It involves deploying a large number of antennas at the base station to serve multiple users simultaneously. Optimizing massive MIMO beamforming using machine learning techniques can further enhance the performance and efficiency of the system.

Machine learning (ML) algorithms can be leveraged to automatically learn and adapt the beamforming parameters based on the network conditions, channel characteristics, and user requirements. Joint beamforming refers to the optimization of beamforming parameters across multiple base stations in a coordinated manner. ML can be applied to enhance joint beamforming by leveraging its capabilities in analyzing complex data patterns, optimizing system performance, and adapting to dynamic environments.

Adaptive cell association refers to the process of dynamically assigning mobile devices to appropriate base stations or cells in a cellular network based on the current network conditions and user requirements. ML techniques can be employed to optimize adaptive cell association by leveraging historical data, real-time observations, and learned models. Load balancing in mobile communication involves distributing the traffic load across different base stations or cells to ensure efficient resource utilization and maintain the quality of service for users. ML techniques can be applied to optimize load balancing by analyzing network data, predicting traffic patterns, and making intelligent decisions. This chapter provides a brief summary of the techniques that are being used in the allocation of resources in mobile communication using ML.

8.2   RESOURCE ALLOCATION USING ML

In mobile communication, resource allocation refers to the assignment and management of constrained network resources including frequency spectrum, time slots, power, and bandwidth to effectively and efficiently service customers. Choosing how to distribute these resources depends on a number of variables, including user needs, network conditions, and quality of service (QoS) requirements. It is difficult to balance trade-offs between many goals, such as increasing system capacity, reducing interference, guaranteeing fairness, and QoS.

To allocate resources efficiently, a variety of optimization techniques are employed, including mathematical programming, game theory, and heuristic algorithms. These methods leverage performance measurements, user needs, and real-time network conditions to make wise resource allocation choices. By examining network data, user requests, and system performance measures, ML techniques can be used to optimize resource allocation decisions. It takes relevant training data, precise models, and appropriate learning algorithms to implement ML for resource allocation. When incorporating ML into resource allocation frameworks, it's crucial to take communication overhead, computational complexity, and system stability into account as well [7]. ML can help mobile communication networks allocate resources more effectively and efficiently by offering insights, forecasts, and optimization capabilities.

8.2.1   Key Aspects of Resource Allocation in Mobile Communication

8.2.1.1   Frequency Allocation

In wireless communication, the frequency spectrum is a valuable and finite resource. To prevent interference and maximize spectral efficiency, resource allocation algorithms decide how to split the available frequency bands among multiple users or cells. To maximize frequency allocation, strategies like dynamic frequency allocation, interference coordination, and frequency reuse are used.

8.2.1.2   Time Slot Allocation

Different users are given time slots in time-division multiple access systems to transmit and receive data. Based on users' traffic demands, channel conditions, and priority levels, resource allocation algorithms choose how to distribute time slots to users. Effective time slot distribution enables equitable resource deployment and reduces interference and collisions.

8.2.1.3   Power Allocation

Power allocation involves determining the transmit power levels for various users or cells. It tries to minimize excessive interference and power consumption while maximizing the signal quality for users. Depending on the channel, user, and traffic conditions, power control algorithms automatically alter the transmit power. In order to get better coverage, lessen interference, and increase energy efficiency, adaptive power allocation is helpful.

8.2.1.4   Bandwidth Allocation

Algorithms for allocating resources take into account things like system capacity, service requirements, and traffic demands to determine how much bandwidth should be given to each user or service. High-demand services are given priority by dynamic bandwidth allocation techniques like adaptive modulation and coding, which also maximize bandwidth use.

8.2.1.5   QoS Considerations

The QoS demands of various services and consumers influence resource allocation decisions.

8.2.2   ML in Resource Allocation

ML can be used in the allocation of resources in mobile communications in the following ways.

8.2.2.1   User Demand Prediction

ML algorithms can forecast user demand for various resources by analyzing previous user data, network usage trends, and contextual data. ML algorithms can predict the anticipated resource needs of users by learning from prior observations. These forecasts can be used to plan ahead and allocate resources effectively to satisfy consumer requests.

8.2.2.2   QoS Optimization

Resource allocation choices can be optimized by ML to guarantee users receive the appropriate level of service; QoS parameters include throughput, latency, packet loss, and dependability. ML algorithms can dynamically prioritize users with high QoS requirements by taking into account elements like user profiles, application requirements, and network conditions. This aids in achieving performance goals and providing users with positive user experiences.

8.2.2.3   Multiobjective Optimization

The difficulty of optimizing numerous conflicting objectives in resource allocation can be handled using ML approaches. ML algorithms can learn to balance trade-offs between many objectives, such as increasing system capacity, reducing interference, and guaranteeing user fairness by defining the resource allocation problem as a multiobjective optimization problem. This enables more flexible and effective judgments on the allocation of resources.

8.2.2.4   Dynamic Resource Allocation

Based on in-the-moment observations and shifting network conditions, ML models are capable of allocating resources in an adaptive manner in response to changes in traffic load, channel conditions, and user demands by continuously analyzing network data, performance indicators, and user behavior. This adaptability aids in maintaining system performance under a variety of circumstances and optimizing resource use.

8.2.2.5 *Reinforcement Learning for Resource Allocation*

Resource allocation strategies can be learned using reinforcement learning algorithms through interactions with the environment and feedback on network performance. The algorithms can maximize system-level performance indicators like throughput, energy efficiency, or user happiness. Reinforcement learning can optimize resource utilization by adapting to shifting network dynamics over time. Algorithms and tactics for allocating resources are constantly changing to meet the problems provided by rising user expectations, expanding data traffic, and the introduction of new communication technologies. To further improve resource allocation efficiency, adaptability, and scalability in mobile communication systems, the deployment of cutting-edge technology, such as ML and artificial intelligence (AI), is being investigated.

8.3 MASSIVE MIMO BEAMFORMING OPTIMIZATION USING ML

8.3.1 Massive MIMO Beamforming

Massive MIMO, as discussed, refers to arrays of hundreds of antennas intended to make possible simultaneous data transmission among large numbers of devices. The process recognizes that in actual systems, environmental noise is filtered out of data being transferred from an antenna to a user terminal and vice versa; the signal may reflect off of structures and other obstructions, and these reflections will affect the signal's arrival time, attenuation, and direction [8]. Even a direct line of sight between the antenna and the user terminal may not exist. It appears that these indirect transmission routes can be used as a positive force [9].

8.3.2 ML Approaches to Optimizing Massive MIMO Beamforming

A number of approaches have been developed to optimize massive MIMO beamforming using the ML algorithms and techniques.

8.3.2.1 *Channel Estimation*

For beamforming optimization, precise channel estimation is essential. The accuracy of channel estimates can be increased by learning the channel characteristics from training data using ML methods like deep learning or reinforcement learning.

8.3.2.2 *Beam Selection and Scheduling*

In order to schedule the transmission resources and to choose the best beams depending on user location, channel conditions, and QoS needs, ML methods can be used. Policies that maximize system performance measures, such as throughput or energy efficiency, can be learned using reinforcement learning.

8.3.2.3 Interference Management

In huge MIMO systems, interference is a significant problem. By modifying the beamforming weights, ML techniques can be used to intelligently reduce interference. Algorithms for reinforcement learning or online learning can be used to dynamically change the beamforming weights in order to reduce interference and improve user signal quality.

8.3.2.4 User Grouping and Precoding

To improve user grouping and precoding techniques, machine learning algorithms can be employed to analyze user behavior and attributes. The system can maximize the performance for each user group by grouping users with comparable channel conditions or traffic patterns in order to optimize the beamforming parameters.

8.3.2.5 Resource Allocation

For huge MIMO systems, ML methods can help with the allocation of radio resources including time, frequency, and power. These algorithms can forecast resource demands and assign them effectively, taking into account elements like user requirements, channel conditions, and system capacity. They can do this by learning from previous data and real-time observations.

It's crucial to remember that ML-based optimization methods for massive MIMO beamforming call for the use of adequate learning algorithms, accurate models, and sufficient training datasets [10]. Massive MIMO systems bring difficulties for ML integration in terms of computational complexity, accessibility to training data, and system stability. However, massive MIMO beamforming in mobile communication systems has the potential to be more adaptable, effective, and efficient thanks to ML.

8.4 JOINT BEAMFORMING USING ML

Joint beamforming, used to enhance the performance of wireless communication systems, is the coordinated transmission of signals from various base stations (or access points). To improve signal quality, expand coverage, reduce interference, and meet other system goals, beamforming weighs multiple base stations. It entails collaboratively optimizing beamforming weights while taking into account variables including channel circumstances, user locations, interference levels, and system objectives [11]. For effective joint beamforming in real mobile communication networks, sophisticated optimization algorithms, ML methods, and distributed coordination protocols are used. The objective is to improve the user experience while maximizing system capacity, coverage, spectral efficiency, and overall performance.

8.4.1   Outline of Joint Beamforming in Mobile Communications

Here we give an overview of joint beamforming in the field of wireless mobile communication.

8.4.1.1   Cooperative Multicell Beamforming

In cellular networks, nearby base stations can collaborate to enhance system performance by collaboratively optimizing their beamforming weights via techniques like joint transmission or coordinated beamforming. Base stations can coordinate their transmissions to enhance signal quality and reduce interference, leading to an increase in capacity and coverage, by sharing channel state information and coordinating beamforming tactics.

8.4.1.2   Interference Alignment

Interference alignment seeks to align interference signals so that they share a common subspace with the intended signal while occupying separate subspaces. This enables the receiver to separate the desired signal from interference and consider the latter as noise. By coordinating the transmit beamforming weights of nearby base stations, interference alignment can be accomplished through joint beamforming optimization.

8.4.1.3   Multiuser MIMO Beamforming

Multiuser MIMO (MU-MIMO) is a method that uses beamforming to provide simultaneous transmission to numerous users. In collaborative beamforming for MU-MIMO, base stations work together to maximize the sum rate or reduce user interference by adjusting their beamforming weights. MU interference can be decreased, resulting in increased spectral efficiency and overall system performance, by coordinating the transmission method and collaboratively optimizing beamforming weights [12].

8.4.1.4   Non-Orthogonal Multiple Access (NOMA)

Using varying power levels and superposition coding, NOMA enables several users to share the same time and frequency resources. Designing beamforming weights specifically for each user in NOMA systems allows for the separation of their signals at the receiver side. Interference between NOMA users can be reduced, resulting in increased capacity and user fairness, by jointly optimizing beamforming weights across many base stations.

8.4.1.5   Hybrid Beamforming

In mmWave and huge MIMO systems, hybrid beamforming is a technology that blends analog and digital beamforming to minimize complexity and power usage. Designing the best analog beamforming weights at the antenna level and the best digital beamforming weights at the baseband processor level is known as joint beamforming optimization in hybrid systems. While respecting hardware limits, system performance can be increased by concurrently optimizing analog and digital beamforming.

8.4.2   ML in Joint Beamforming

ML has a number of applications in joint beamforming.

8.4.2.1   Cooperative Beamforming

ML models can improve the beamforming parameters across various base stations to improve system capacity, coverage, and interference control by examining the channel conditions, traffic patterns, and user needs. To develop cooperative beamforming techniques that improve system-wide performance measures, distributed learning or reinforcement learning algorithms can be used.

8.4.2.2   Cross-Cell Interference Management

Cross-cell interference can significantly lower system performance in dense networks. To reduce interference and boost signal quality, beamforming weights can be simultaneously optimized across neighboring cells using ML. Reinforcement learning or deep learning algorithms are able to adaptively change the beamforming settings to reduce cross-cell interference while increasing user throughput [13].

8.4.2.3   Dynamic Beamforming Adaptation

In response to changing network conditions, beamforming parameters can be continually adjusted using ML techniques. ML algorithms can forecast the best beamforming configurations for various scenarios by examining real-time channel status information, traffic trends, and user mobility. This adaptive beamforming can help to increase resource efficiency, preserve signal quality, and optimize system performance.

8.4.2.4   Learning-Based Resource Allocation

Resource management across several base stations is necessary for joint beamforming. The distribution of radio resources, such as time, frequency, and power across base stations, can be optimized using ML approaches. Resource needs, user traffic patterns, and channel conditions can all be predicted using ML models that can learn from both previous data and current observations.

8.4.2.5 Learning-Based Coordination

ML models can discover coordination techniques to improve beamforming parameters, handover selections, and interference control across base stations by examining historical data and network performance indicators. Coordination strategies that maximize system performance can be learned using distributed learning or reinforcement learning methods. It takes the right training data sets, precise models, and efficient learning algorithms to apply ML to joint beamforming. When incorporating ML into joint beamforming frameworks, it is further crucial to take computing complexity, communication overhead, and system stability into account. The performance and effectiveness of mobile communication systems can be improved by collaborative beamforming with the help of ML.

8.5 ADAPTIVE CELL ASSOCIATION AND LOAD BALANCING USING ML

8.5.1 Adaptive Cell Association Using ML

Adaptive cell association in the context of mobile communication refers to the dynamic assignment of mobile devices to various base stations (cells) based on current network conditions and user needs. To enhance network performance, coverage, capacity, and user experience, it is important to align mobile devices and base stations optimally. For effective resource management, increased coverage, load balancing, and improved user experience in mobile communication, adaptive cell association is crucial. To dynamically alter the relationship between mobile devices and base stations requires real-time network state monitoring, reliable handover mechanisms, and clever algorithms. Mobile communication networks can increase network capacity, handle traffic changes better, and offer smooth connectivity to consumers by optimizing cell association.

8.5.1.1 Synopsis of Adaptive Cell Association

The compendium of the adaptive cell association in the field of wireless communication can be described as follows.

8.5.1.1.1 Signal Quality and Coverage

The signal strength and coverage offered by various base stations are taken into consideration during adaptive cell association. Devices are connected to the base station that provides the area's strongest and most dependable signal. Devices can maintain optimal connectivity and prevent signal degradation or coverage gaps by dynamically modifying the cell association.

8.5.1.1.2   Load Balancing

Adaptive cell association aids in distributing the traffic load among base stations. The network can efficiently use its resources and avoid congestion in some cells by evenly distributing mobile devices among several base stations. As a result, there is better resource allocation, less interference, and greater overall network performance.

8.5.1.1.3   Mobility Management

For controlling the movement of mobile devices, adaptive cell association is essential. The smooth transfer of gadgets from one cell to another while they are in motion is required. In order to maintain a seamless changeover procedure and uninterrupted connectivity as devices travel between coverage regions, the network can initiate cell re-association by monitoring the signal strength and quality.

8.5.1.1.4   QoS Considerations

QoS specifications and user preferences are taken into account during adaptive cell association. Depending on the resources that are available, the level of congestion, and the network conditions, different base stations deliver different levels of QoS. Adaptive cell association algorithms take into account user-specific QoS needs and pair devices with base stations that can fulfill them.

8.5.1.1.5   Dynamic Network Conditions

Adaptive cell association changes with the network's varying conditions. Congestion levels, interference levels, and channel quality might change over time. Adaptive cell association algorithms can change how mobile devices are associated to maximize performance based on the current network conditions by continuously monitoring these aspects.

8.5.1.1.6   ML and Optimization

To enhance adaptive cell association, ML and optimization methods can be used. To choose the best cell association strategy, these algorithms consider user behavior, historical network data, and real-time observations. ML algorithms are able to forecast the best cell association for a specific device or network situation by learning from prior observations.

8.5.1.2 ML in Adaptive Cell Association

ML can be used in adaptive cell association for mobile communication in a number of ways.

8.5.1.2.1 Learning-Based Prediction

In order to forecast the best base station for a specific mobile device at a specific time, ML algorithms can examine historical data, user movement patterns, and network performance parameters. ML can predict which base station will likely offer the highest level of service quality for a given device in real time by training models on previous user behavior and network conditions.

8.5.1.2.2 User Experience Optimization

Signal strength, throughput, latency, and user preferences are just a few of the variables that can be considered when using ML to enhance the user experience. User satisfaction will increase as a result of ML models' adaptive assignment of mobile devices to base stations that offer higher service quality.

8.5.1.2.3 Context-Aware Decision-Making

Intelligent cell association decisions can be made by ML algorithms by using contextual data like location, user behavior, and network load. ML algorithms can adaptively choose the base station that minimizes interference, maximizes resource utilization, and enhances overall system performance by analyzing various contextual factors.

8.5.1.2.4 Reinforcement Learning for Cell Association

Through trial-and-error interactions with the environment, reinforcement learning algorithms can be used to learn cell association regulations. The algorithm's decision-making process can be modified in response to feedback on the connection's quality, the state of the network, and the user experience. The reinforcement learning model can eventually learn to select cell associations based on incentives or feedback signals that maximize system performance.

8.5.1.2.5 Online Learning and Adaptation

ML algorithms are able to continuously learn from and adjust to changing user behavior and network conditions. The models may update their knowledge based on in-the-moment observations by using online learning approaches, enabling dynamic modifications in cell association decisions. This flexibility is essential in settings where network conditions or user mobility habits change over time.

It should be noted that applying ML for adaptive cell association calls for pertinent training data, precise models, and appropriate learning techniques. Additionally, when incorporating ML into adaptive cell association frameworks, considerations for processing resources, communication overhead, and system stability must be made [14]. In cellular communication systems, ML has the ability to optimize cell association choices, improve user experience, and boost overall network performance.

8.5.2   Load Balancing Using ML

The process of dispersing traffic and resource usage across various base stations or cells in a cellular network is known as load balancing in the context of mobile communications. The goal is to enhance system performance, maximize the usage of network resources, and improve user experience. In mobile communications, efficient load balancing helps to improve network capacity, decrease congestion, and maximize the usage of network resources. It makes it possible to utilize network infrastructure effectively, reduces service interruptions, and guarantees that resources are distributed fairly among users and cells. The continual improvement of load-balancing techniques in mobile communication networks is facilitated by ongoing developments in load-balancing algorithms, data analytics, and ML [15].

8.5.2.1   Traffic Monitoring and Measurement

The first step in load balancing is to track and gauge the volume of traffic in various base stations or cells. In order to determine how the network is being loaded, various factors are tracked, including user count, data volume, call volume, and signaling load.

8.5.2.2   Load Assessment and Prediction

The current load on each base station is determined using load assessment techniques, which examine the measured traffic data. The future load can also be predicted using load prediction algorithms based on historical data, network trends, or user behavior patterns.

8.5.2.3   Load Redistribution

The ideal traffic allocation among the base stations is decided by load-balancing algorithms. These algorithms determine which customers or services should be shifted from densely populated cells to sparsely populated cells based on load assessment and prediction. Handover, cell reselection, or traffic unloading are just a few of the processes that can be used to redistribute the load.

8.5.2.4   Cell Range Expansion (CRE)

Increasing the transmit power levels of underutilized cells allows for cell range expansion, which increases the coverage area of such cells. Devices in adjacent cells can be offloaded to cells with low traffic loads, balancing the network's total load, by extending the range of these cells. Congestion is avoided and total network performance is maximized by ensuring that resources are distributed effectively to meet the demands of each cell [16]

8.5.2.5   Dynamic Resource Allocation

Dynamically assigning resources, such as frequency channels, time slots, or power, to meet traffic demands is known as load balancing. Resource allocation algorithms take into account things like traffic patterns, QoS needs, interference levels, and system capacity to maximize the use of the resources that are already available.

8.5.2.6   Mobility Management

Mobility management and load balancing are closely linked concepts. Load balancing algorithms guarantee a smooth handover procedure when a user switches between cells by moving the user's connection from a cell that is overloaded to one that is underloaded. This prevents congestion in particular locations and preserves a balanced distribution of users throughout cells.

8.5.2.7   Proactive and Reactive Approaches

Both proactive and reactive methods of load balancing are available. Proactive load balancing solutions take proactive measures to balance the load in advance of anticipated load situations. Real-time traffic distribution adjustments are made through reactive load balancing techniques in response to sudden load imbalances.

8.5.2.8   ML and Optimization

Clustering algorithms and reinforcement learning are two examples of ML techniques that can be used to analyze past data, forecast traffic patterns, and improve load-balancing choices. Additionally, load-balancing issues can be resolved and the best traffic redistribution plan can be discovered using optimization techniques. It's crucial to remember that applying ML for load balancing calls for pertinent training data, precise models, and suitable learning methods [17].

When incorporating ML into load-balancing frameworks, it is also necessary to take into account computing complexity, communication overhead, and system stability. To maintain a balanced load and improve the efficiency of mobile communication networks, ML can offer insights, predictions, and optimization skills.

REFERENCES

[1] Le Liang, Y. H., Yu, G., and Li, Y. G., "Deep-learning-based wireless resource allocation with application to vehicular networks," Proceedings of the IEEE, vol. 108, no. 2, (2020), pp. 341–356.

[2] Melhem, S. B., Kaushik, A., et al. "Machine learning for resource allocation in mobile broadband networks," Communication Networks and Service Management in the Era of Artificial Intelligence and Machine Learning, (2021), pp. 123–146.

[3] Hui, H., "Intelligent resource allocation method for wireless communication networks based on deep learning techniques," Journal of Sensors, (2021), pp. 1–12.

[4] Shrivastava, S., & Thakur, R. (2024, January). "Resource Allocation Optimization in Local and Wide Area Networks through Social Network Integration." In Proceedings of the 25th International Conference on Distributed Computing and Networking (pp. 254–255).

[5] Sangeetha, S., Logeshwaran, J., Faheem, M., Kannadasan, R., Sundararaju, S., & Vijayaraja, L. (2024). "Smart performance optimization of energy-aware scheduling model for resource sharing in 5G green communication systems." The Journal of Engineering, 2024(2), e12358.

[6] Sarah, A., & Nencioni, G. "Resource allocation for cost minimization of a slice broker in a 5G-MEC scenario." Computer Communications, 213, (2024), pp. 331–344.

[7] Nazir, M., Sabah, A., et al. "Power and resource allocation in wireless communication network," Wireless Personal Communications, vol. 119, no. 4, (2021), pp. 3529–3552.

[8] Carrera, D. F., et al., "Comparative study of channel estimators for massive MIMO 5G NR systems." IET Communications, vol. 14, no. 7, (2020), pp. 1175–1184.

[9] Taşpınar, N., Ergeç, A., & Gül, B. K. "Pilot Tones Design for Channel Estimation Using Elephant Herding Optimization Algorithm in Massive MIMO Systems." Wireless Personal Communications, (2024), pp. 1–18.

[10] Maksymyuk, T., Gazda, J., et al., "Deep learning based massive MIMO beamforming for 5G mobile network," IEEE 4th International Symposium on Wireless Systems within the International Conferences on Intelligent Data Acquisition and Advanced Computing Systems (IDAACS-SWS), (2018), pp. 241–244.

[11] Darsena, D., Gelli, G., and Verde, F., "Beamforming and precoding techniques," arXiv preprint arXiv:2004.05079, (2020).

[12] Dilli, R., "Hybrid beamforming in 5G NR networks using multi user massive MIMO at FR2 frequency bands," Wireless Personal Communications, (2022), pp. 1–33.

[13] Mismar, F. B., Evans, B. L., and Alkhateeb, A., "Deep reinforcement learning for 5G networks: Joint beamforming, power control, and interference coordination," IEEE Transactions on Communications, vol. 68, no. 3, (2019), pp. 1581–1592.

[14] Elsherif, A. R., Chen, W. P., et al., "Adaptive resource allocation for interference management in small cell networks," IEEE Transactions on Communications, vol. 63, no. 6, (2015), pp. 2107–2125.

[15] Gures, E., and Yazici, I., "A comparative study of machine learning-based load balancing in high-speed," Alexandria Engineering Journal, vol. 72, (2023), pp. 635–647.

[16] Hasan, M. K., and Chuah, T. C., "Constriction factor particle swarm optimization based load balancing and cell association for 5G heterogeneous networks," Computer Communications, vol. 180, (2021), pp. 328–337.

[17] Girma, S. T., and Abebe, A., G., "Mobility load balancing in cellular system with multicriteria handoff algorithm," Advances in Fuzzy Systems, (2017), pp. 1–8.

Chapter 9

Reciprocated Bayesian–Rnn Classifier-Based Mode Switching and Mobility Management in Mobile Networks

*Shashi Kant Gupta, Anchal Pathak,
Sultanuddin SJ, and Nupur Soni*

9.1 INTRODUCTION

In an era where smartphones have become nearly ubiquitous and 5G technology continues to proliferate, the demand for efficient mobile management solutions has reached new heights. The latest developments in this field have focused on optimizing the performance of 5G networks through advanced management techniques and rigorous performance analysis. With the advent of technologies like Reciprocated Bayesian–RNN Classifier-Based Mode Switching and Mobility Management, significant strides have been made towards enhancing the efficiency and effectiveness of 5G mobile management. Through sophisticated algorithms and machine learning capabilities, these solutions enable dynamic adjustment of device operation modes and streamline mobility management processes. Moreover, comprehensive performance analysis has highlighted the tangible benefits of such approaches, showcasing improvements in network throughput, latency reduction, and overall user experience. As 5G technology continues to evolve, these innovations promise to further elevate the capabilities of mobile networks, ensuring seamless connectivity and optimal resource utilization for users worldwide **(Banafaa et al., 2024; Shabbir et al., 2024, Krishnan et al., 2024)**.

As a result, the mobile sector is seen as a stream with significant technological and commercial rivalry, in which quality of system (QoS) is a critical aspect. Quality is determined based on several variables, including feasibility, dependability, and the speed with which services are delivered. Mobility management is critical for seamless communication as users travel across networks, and with its unique characteristics, bandwidth is a valuable resource in this process; low bandwidth can cause network overload, service breakdown, and slow communications. Researchers determined that tracking mobile users as they travel consumes 40% of signaling traffic; in 2G, 3G, and 4G networks, tracking users created bottlenecks, but methods are being optimized. In resolving the mobility management problem, metaheuristics have been a popular approach among academics **(Singh and Prasad 2021)**.

Mobility management is a fundamental feature that allows mobile terminals to use network services when traveling without experiencing service interruptions.

DOI: 10.1201/9781003306290-9

The rapid growth in mobile broadband traffic has resulted in a large need for increased network capacity and speed. In today's network design, mobility management is a major challenge. For smooth mobility, a unified framework that makes use of current radio access technology and incorporates capabilities like a multimode terminal interface is required. Mobility management is concerned with the control of a user's mobility in a diversity of computer environments (**Zhang et al. 2019**). Figure 9.1 illustrates the general framework for mobility management.

Various protocols offer mobility at different layers of the protocol stack. It's a collaborative effort that's aided by a slew of supporting modules and agents. In the next part, we'll go over the fundamental structure of mobility, which explains how these entities are connected, what challenges the generic design faces, and how they've been solved using a hierarchical framework.

In the realm of modern cellular networks, 5G—the fifth generation of mobile networks—is gaining traction, already operational in many developed nations. The new technology has much more desired characteristics than current communication technologies. There are a few problems and complexities associated with putting 5G into real-time activity (**Barua 2018**).

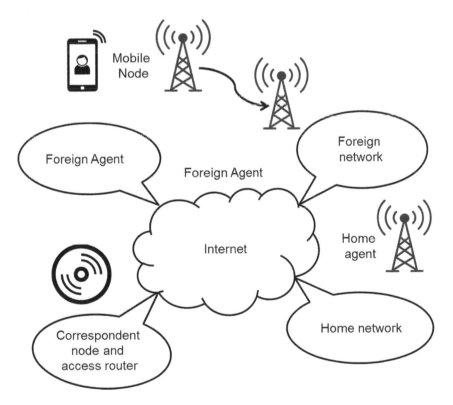

Figure 9.1 General Framework for Mobility Management.

Academic circles, companies, governmental associations, and standards agencies collaborated to arrive at features that the new generation of mobile networks should encompass, which included device-centric communication supported by massive multiple-input/multiple-output (M-MIMO) antenna arrays. One of the most important concerns in 5G networks is efficient resource allocation; in comparison with current 4G/LTE networks, the essential duty of next-generation 5G wireless networks is to provide increased data speeds (on the order of gigabytes per second) with very low latency **(Dahi et al. 2022)**.

Cellular genetic algorithms (CGA) are evolutionary algorithms that methodically organize the inhabitants in a network, with every node representing a person and the communications between them regulated by the neighborhood. In addition to CGAs, there are a variety of additional algorithms that are often used to resolve issues such as efficient mobility management; defining and fine tuning the parameters determines the output of the models. As a result, fresh loop tuning is required to restore the problem's efficacy without the need for an external agent. Simultaneously, constraint change adaptation may affect multidimensional algorithms with more hidden-tunable parameters and sophisticated calculations than static algorithms. In addition, deep learning methods include deterministic, which uses offline deterministic rules; adaptive, which uses algorithm feedback; and self-adaptive parameters, which are developed with a problem solution in mind **(Dahi et al. 2018)**.

The goal of this study was to develop a mode-switching method for mobile networks that will help with mobility management. Mobile networks are established to build routes so that communication may begin. The optimum routing algorithm model is used to find routes once they have been formed. The communication is then initiated when the optimum route has been selected, followed by active form switching, in which the system shifts the form of a message from mobile network mode to user mode to achieve a reliable connection. For mode selection, these switching factors are input into a reciprocated Bayesian-recurrent neural network (RB-RNN). Mobility management is performed after mode switching for route maintenance **(Sapkale and Kolekar 2021)**. For RB-RNN mode switching, the network shifts communication modes from mobile to user mode depending on a switching factor that takes into account elements such as consumer equipment or objective node signal strength, bandwidth usage, power of user mode or mobile mode, the latency of consumer mode or mobile mode, and network usage of consumer mode or mobile mode **(Gao et al. 2019)**.

Furthermore, with the exponential growth of data consumption and the proliferation of Internet of Things (IoT) devices, the need for efficient 5G mobile management has become paramount. The latest advancements in this arena have also addressed the challenges posed by the diverse range of connected devices and applications. Advanced techniques such as network slicing and edge computing have been leveraged to tailor mobile management strategies to the specific requirements of different use cases and applications. By dynamically allocating resources and optimizing network configurations, these solutions can effectively support the diverse needs of modern digital ecosystems. As a result, 5G networks are not only capable of delivering high-speed connectivity to smartphones but also facilitating the seamless integration of a myriad of IoT devices and emerging technologies, thereby laying the groundwork for a truly connected and intelligent future (**Mu, & Antwi-Afari, 2024; Lilhore, & Simaiya, 2024**).

9.2   PROBLEM STATEMENT

The aim of this paper is to establish an efficient method of mobility manage-ment as users travel across mobile networks. Challenges in mobility management include double interconnections, Internet backbone variability, linked drones, hyper systems, ineffective control processes, and incomplete enhancement as well as those we have already discussed.

9.3   PROPOSED WORK

In this chapter, we present a mobility management system based on RB-RNN. The general depiction of the planned work is illustrated in Figure 9.2.

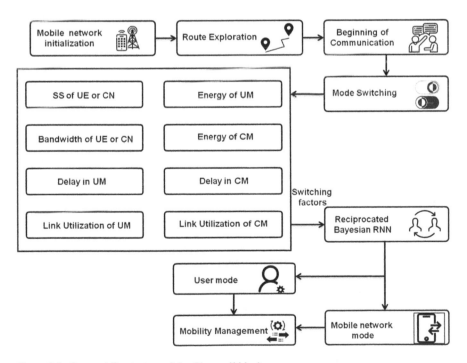

Figure 9.2 General Depiction of the Planned Work.

9.3.1   Mobile Network Initialization

$B = (B_1, B_2 \ldots B_p)$ represents a collection of substance routers in the network, and $U = (B, H)$ symbolizes a group of bidirectional connections to attach plans in the network. Assume that $V = (V_1 V_2 \ldots V_s)$ is a set of content objects that may be found on the network. The majority of these items are spread in network servers, which are inextricably linked to border routers. The purpose is to achieve the lowest possible network latency by addressing the problem of network caches with limited capacity. Routing is optimized following this formula:

$$\min_{m,n} : \sum_{i=1}^{P}\sum_{j=1}^{T} U_i^j \sum_{r=1}^{P} t_{i,r,j} V_{i,r,j} \tag{1.1}$$

where U_i^j represents the request rate for item v_j at node i and $s_{i,r,j}$ represents the distance traveled by node i to get content object v_j as of node r. $n_{i,r,j}$ is a constituent that has an assessment of 1 if node i downloads a duplicate of the contented point j as of node r, which is written as

$$\sum_{r=1}^{P} v_{i,r,j} = 1, \ \forall i, j \tag{1.2}$$

$$v_{i,r,j} \leq Z_{r,j}, \ \forall i, r, j \tag{1.3}$$

$$\sum_{j=1}^{T} W^j Z_{i,j} \leq I_i, \ \forall i \tag{1.4}$$

where I_i denotes the highest accumulation mass at the router i and z^j denotes the contented object V_j's size. If node x stores a reproduction of constituent j, $Z_{i,j} = 1$; otherwise, $Z_{i,j} = 0$. This is mathematically written as

$$V_{i,r,j} \in \{0.1\} \ and \ Z_{i,j} \in \{0,1\}, \forall i, r, j \tag{1.5}$$

9.3.2   Route Exploration

The route finding is carried out here with the help of an optimum direction-finding algorithm. Initially, the manager monitors the system and sends out data over a set phase of time. It can count the number of consumer needs and conditions of real-time loading based on observation. When a mobile consumer needs content, the need and substance name is wrapped in a wrapper and transmitted to the network's border appliance. The regulator then uses the gathered data to determine the best route with the minimum network charge between the suppliers, which grasps the content of consumer needs.

Following that, the route-finding system determines the state of in-network caches depending on the number of consumer needs and accumulation substitution rules. The organizer produces the ideal routing route and improves the cache state depending on the qualities of the Big Data platform. The application initially alerts the organizer about the kinds and quantity of possessions required in the request packet to meet the demands of the application. When the demand package is established, the organizer uses the Big Data dispensation stage to analyze the resource information in the package and subsequently distributes possessions depending on the relevant order regularly. Finally, the virtual network is created, and the optimum routing algorithm replica is used to decrease network latency in underregulated settings by examining the storage space and contented condition of the virtual network.

9.3.3   Mode Switching

The naive Bayes classifier is used to represent the dynamic mode switching process, in which the switching variables are fed into the classifier to pick the virtual network (VN) or dynamic network (DN) for the dependable message. The user-centric form is smaller dependent on system communications, and the affiliated improved node allows for further scrutiny of performance metrics. The Bayesian theorem is used to create a replica for grouping and categorization called reciprocated Bayesian. The per-class probability is estimated using this model, which assumes that the qualities are provisionally self-governing.

It works similarly to predict the likelihood of an unusual program based on different qualities. The naive Bayes model is typically used to solve issues involving numerous classes. It provides several benefits, for instance not requiring as much training data and being simple to set up. It can also be used to make predictions in real-time and deal with both discrete and continuous data. The following is a quick example of the suggested mode-switching process:

* The Voice Function (VF) initiates a call, and the associated enhanced node overtakes it.
* The upgraded node calculates the cell and caller's locations, as well as the target node's position.
* For each VF and Destination Mobile (DM), which is the intended mobile user, the upgraded node calculates switching variables such as signal strength, bandwidth, energy consumption, latency, and link usage.
* The naive Bayes classifier uses the collected switching factors to decide whether the user should operate in Voice Network (VN) or Data Network (DN) mode.
* The improved node in VN assesses each mobile user's Quality of Service (QoS) parameter. If the QoS values are less than the threshold value, it attaches a relay node to improve network performance.
* The system's presentation is prohibited by the enhanced node, which uses user-centric pairing to provide sufficient QoS.

The network keeps track of performance and adjusts various activities to provide better service to mobile consumers. There are two subordinate stages in this module. The network first adjusts the visitor in order and investigates it. Then, when switching between VN and DN, a judgment is made by the naive Bayes classifier for controlling modes based on presentation measurements. Signal strength, bandwidth, energy, latency, and connection usage are calculated to enhance mode switching performance.

9.3.4   Effectiveness Metrics for Assessing Mobile Users

In mobile networks, a mobile node may communicate with another device through the mobile network's infrastructure, such as enhanced nodes. In the cellular network, this strategy is known as device-centric or user-centric communication. The user-centric pairings are altered to increase network arrangement using switching factors such as signal intensity, bandwidth, energy, latency, and link usage.

9.3.4.1   *Signal Intensity*

The output power established by the spreader using an orientation transmitter at a specific distance from the transmitting transmitter is referred to as signal strength.

9.3.4.2   *Bandwidth*

The maximum pace of data transport in a particular route is referred to as bandwidth in computing. The bandwidth usage between the resource and purpose nodes is expressed as follows:

$$A = \sum_{y=o}^{w} \eta_f \rho_g \prod_l^x AD + \sum_{y=o}^{w/2} h_f \eta_f \rho_g \prod_l^x AD \tag{1.6}$$

B stands for bandwidth allotment, while D stands for the service area.

9.3.4.3   *Energy*

The energy consumption of a mobile network is classified into two parts: networking power and device power. Pack bond length, networking refresh power, and data packet power are the three components of network energy that are centered here. Hop count and movement element of use, which are estimated using DN and VN, determine these components. In DN, energy usage is expressed as

$$\varepsilon_{DN} = \prod_l^x \sum_{f=1}^{\gamma} \sum_{g=1}^{v} (\alpha + \beta + \gamma) + \prod_l^x \sum_{f=1}^{\gamma} \sum_{g=1}^{V} \chi \tag{1.7}$$

where α is for unit energy, γ represents hop count $(1 \leq f \leq \gamma)$, I is signaling cost, V stands for user mobility factor $(1 \leq g \leq v)$, and V stands for network update energy. VN's energy usage is calculated as follows:

$$\varepsilon_{VN} = \prod_l^x \sum_{f=1}^{\gamma} \sum_{g=1}^{v} (\alpha + \beta + \gamma) + \prod_l^x \sum_{g=1}^{V} \chi \tag{1.8}$$

9.3.4.4   Latency

The manner of communication is determined by the network's performance, and the communication delay is determined by dispensation and proliferation time. The resulting DN stoppage is calculated as follows:

$$P_{DN} = Vh_f \left(\prod_l \sum_{f=1}^{x} \sum_{y=0}^{\gamma} \frac{\partial Y}{\partial} \rho + \pi_x^l \sum_{y=0}^{\frac{\partial Y}{\partial}} \rho \right) \tag{1.9}$$

where V denotes the movement coefficient and D denotes the communication latency.

The VN delay is determined by

$$P_{VN} = Vh_f \left(\prod_l \sum_{f=1}^{x} \sum_{y=0}^{\Psi} \frac{\partial Y}{\partial} \rho + \pi_x^l \sum_{y=0}^{\frac{\partial Y}{\partial}} \rho \right) \tag{1.10}$$

where N = 2 denotes a two-hop communication system

9.3.4.5   Link Usage

The average traffic in a given connection is measured as a proportion of the total link capacity to determine link usage. The efficiency of the connection, which is defined as the ratio of time spent sending data to the total time spent acknowledging data, is a less often used term. Then, in DN, link usage is written as

$$\pi_{DN} = \max_{p_d, A_d} \left(\frac{B_d F_d}{X_d + X_{min}} \right) \tag{1.11}$$

where p_d represents the proportion of resource allocation in DN, A_d represents bandwidth utilization in DN, F_d represents transmission rate in control mode, X_{min} represents the minimal power allocation in DN, X_d represents power owed in DN, and B_d represents energy wasted in DN.

The following is a formula for link usage in VN:

$$\pi_{VN} = \max p_{v}, A_{v} \left(\frac{B_{v} F_{v}}{X_{v} + X_{min}} \right) \tag{1.12}$$

where p_v is the proportion of VN resource allocation, B_v denotes UM bandwidth usage, F_v denotes VN transmission rate, X_{min} denotes the minimum power allocation in VN, X_v denotes the amount of power owed in VN, and A_v denotes the amount of energy dissipated in VN.

9.3.5   Reciprocated Bayesian-RNN Classifier

The RB-RNN classifier uses a set of standard measurements and dynamic mode switching. For practical communication, the network transforms the transceiver from network-centric to user-centric. For analyzing the connectors or mobile mode, five performance measures are fed into the classifier.

The reciprocated Bayesian classifier uses the Bayes theorem to determine the likelihood of data fitting into a certain class given examinations. This replica calculates per-class probability by taking into account provisionally self-governing characteristics. This classifier is easy to construct and does not need any sophisticated iterative parameters and evaluation, making it ideal for hardware implementation. Machine learning, which is often tailored for creating classifications using datasets, is known to be supervised by reciprocated Bayesian. This classifier predicted something based on its prior experience and assumptions. Furthermore, using anonymous data sets, this classifier is successful in generating smart judgments.

RNN helps each of the mobile users that we track, the actual values of section II-A input's characteristics are measured in real-time. A likelihood ratio is calculated for each feature using the likelihood purpose determined in II-B. All the data gathered must be combined in a higher-level administrative stage where probable mistakes are compensated, resulting in a small amount of bogus freight and traffic control. The first task in integrating different features is to determine the probability of a DoS assault by standardizing the attack probability for every attribute:

$$
h_{final} = \frac{h_{bit} + h_{acc} + h_{entr} + h_{Hurst} + h_{delay} + h_{delrate}}{total\ number\ of\ features} \tag{1.13}
$$

The value of h_{final} is between 0 and 1. The comparison of this assessment to a set entrance, which may or may not is based on the effect that the signal attack is predicted to have on the casualty, is then used to determine if the incoming signal is normal or not:

$$
e^+(y,x) = s(x)t^+(x,y) \geq 0 \tag{1.14}
$$

The estimated probabilities are employed as input in an RNN in the second version of the detection process. A feed-forward RNN structure with six inputs, twelve neurons in one hidden layer, and two outputs employed. The probability ratios for the six input characteristics are fed into the inputs, and the output nodes correspond to normal and assault mobility. RNN implementation is provided for the data fusion. Scales, both positive and negative, are calculated using

$$
e^-(y,x) = s(x)t^-(x,y) \geq 0 \tag{1.15}
$$

$$B(x) = \sum_y w_y e^+ (y,x) + \wedge (x) \tag{1.16}$$

$$s(x) - firing\, rate$$

The potential for neuron x is $w_x = \dfrac{C(x)}{s(x)}$

$$C(x) = s(x) + \sum_y w_y e^- (y,x) + \lambda (x) \tag{1.17}$$

$$s(x) = \sum_y e^+ (x,y) + e^- (x,y) \tag{1.18}$$

For comparison, we used a simple detection method that just used the RNN module and the underdone standards for the input characteristics.

9.3.6 Mobility Management

Mobile users in the mobile model are free to shift on any track about the communication infrastructure. If any node changes its position during communication, the associated performance may suffer since it affects the Q. The mobility management model is in charge of controlling user mode and cellular model-based communication, and if it falls below a certain level, communication is restored to the traditional model. Node mobility is a key aspect in achieving cellular communication since it enables users to go in any direction for higher QoS. In specific coverage zones, the user's starting position in the 3D plane is considered to be $p_1\left(l_1, \lambda_1, u_1\right)$, and the current location is assumed to be $p_2\left(l_2, \lambda_2, u_2\right)$ The distance between the starting and current positions is calculated as follows:

$$Dist = |p_1 p_2| \tag{1.19}$$

$$Dist = \sqrt{(l_2 - l_1)^2 + (\lambda_2 - \lambda_1)^2 + (u_2 - u_1)^2} \tag{1.20}$$

9.4 PERFORMANCE ANALYSIS

To measure the performance of the proposed RB-RNN classifier in mode switching and mobility management in mobile networks, we compared it with the performance of convolutional neural network, long short-term memory, linear support vector machine, and naive Bayes according to the following criteria: location management, throughput, routing efficiency, latency, network usage, power, and security.

9.4.1   Location Management

Other nodes wishing to use the mobile host's services must be able to contact it; it must also be accessible to other nodes while moving and retain location transparency. The suggested method's location management range is compared to that of previous approaches. The proposed method shows high accuracy for location management. Figure 9.3 depicts the comparative analysis for location management.

9.4.2 Throughput

The overall data rates transferred across the mobile network in a given period are referred to as the throughput of the system:

$$Through\,put = \frac{Amount\,of\,data\,received}{given\,period} \tag{1.21}$$

Compared with the current approaches, the suggested method has greater throughput. The throughput comparison is depicted in Figure 9.4.

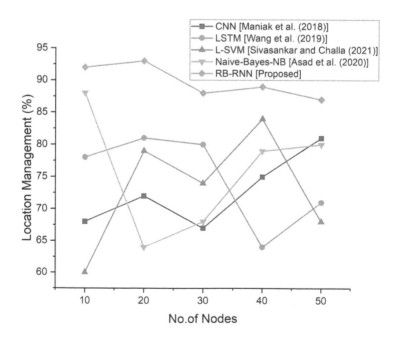

Figure 9.3 Comparative Analysis of Location Management.

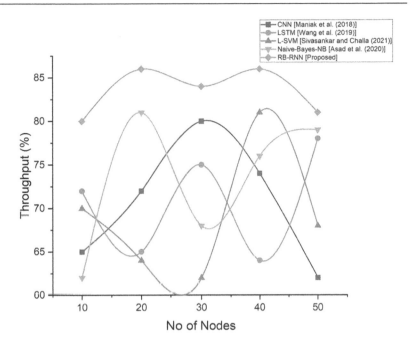

Figure 9.4 Comparative Analysis of Throughput.

9.4.3   Routing Efficiency

Routing protocols apply the same criteria to all nodes in a network, whether they are source nodes or nodes that route user data. The routing protocols in the group give a path to the source for each sensor node. Packets are transmitted with the lowest feasible delay by adopting the quickest route offered by the routing protocol. Compared with the current approaches, the suggested method has greater routing efficiency. The comparison study for routing efficiency is depicts in Figure 9.5.

9.4.4   Latency

Regardless of the attacker, the latency parameter specifies the overall time required to transfer the data. The latency is calculated using the following formula:

$$Latency = \frac{Amount\ of\ data}{Transmission\ speed} \tag{1.22}$$

Compared with the current approaches, the suggested method has low latency. The latency comparison is depicted in Figure 9.6.

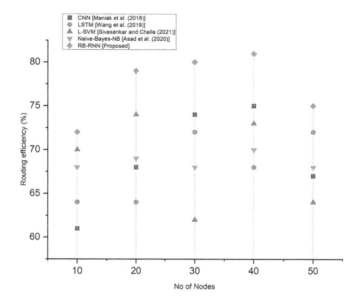

Figure 9.5 Comparative Analysis of Routing Efficiency.

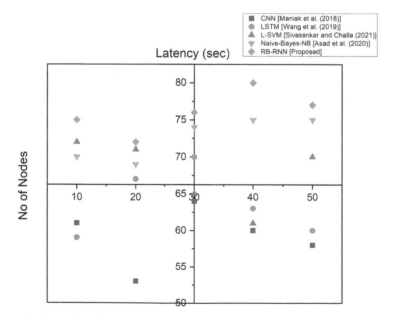

Figure 9.6 Comparative Analysis of Latency.

9.4.5   Network Utilization

The average traffic in a given connection while transferring data is referred to as network utilization. Compared with current approaches, the suggested method has a greater network usage level. The comparison study for network usage is depicted in Figure 9.7.

9.4.6   Power Requirement

The energy necessary to transport data from one cell to another is referred to as power. The suggested approach has a lower power need than the current methods. The power needs comparison is depicted in Figure 9.8.

9.4.7   Security

In a wireless setting, security is a key concern since it requires authentication of the subsequently accessed node in the public networks, which adds overhead during handover. The network should not be burdened with extra security overheads as a result of mobility management systems. Figure 9.9 depicts the comparative analysis of security.

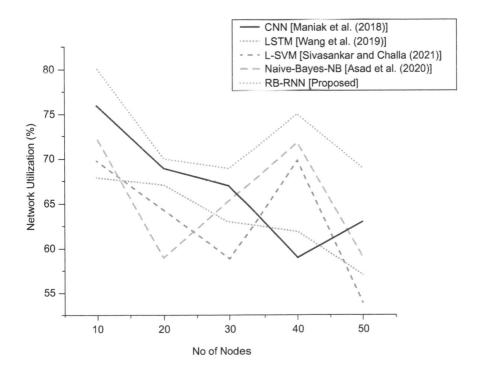

Figure 9.7 Comparative Analysis of Network Utilization.

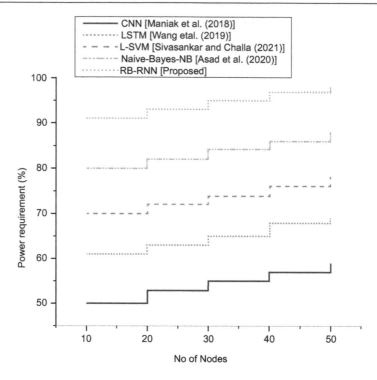

Figure 9.8 Comparative Analysis of Power Requirement.

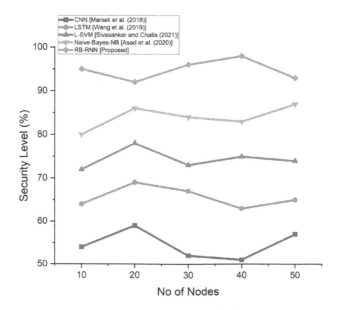

Figure 9.9 Comparative Analysis of Security.

9.5   CONCLUSION

To control mobility in cellular networks, we propose a unique mode-switching strategy. The solution incorporates dynamic quality of system-based mode switching and network mobility control. The network examines the switching factors for each mobile user, which include specific performance-based criteria such as connection usage, bandwidth, latency, energy consumption, and signal strength. We updated the reciprocated Bayesian recurrent neural network (RB-RNN) classifier to convert the user's processing mode from cellular to user-centric. The switching parameters are used as input for the classifier to switch the modes. Mobility management is performed after mode switching for route preservation. Mobility between nodes is an important aspect of mobile network communication since it enables the user to move on any track for better quality of service.

REFERENCES

Asad, S.M., Ansari, S., Ozturk, M., Rais, R.N.B., Dashtipour, K., Hussain, S., Abbasi, Q.H. and Imran, M.A., 2020. Mobility management-based autonomous energy-aware framework using machine learning approach in dense mobile networks. Signals, 1(2), pp. 170–187.

Banafaa, M., Pepeoğlu, Ö., Shayea, I., Alhammadi, A., Shamsan, Z., Razaz, M. A., ... & Al-Sowayan, S. 2024. A Comprehensive Survey on 5G-and-Beyond Networks with UAVs: Applications, Emerging Technologies, Regulatory Aspects, Research Trends and Challenges. IEEE Access.

Barua, S., 2018. D2D communications in 5G mobile cellular networks: We propose and validate a novel approach to mobility management (Doctoral dissertation).

Berrocal-Plaza, V. and Vega-Rodríguez, M.A., 2021. Analysis and comparison of mobility management strategies in public land mobile networks from a multi objective perspective. Journal of Network and Computer Applications, 177, p. 102967.

Dahi, Z.A., Alba, E. and Draa, A., 2018. A stop-and-start adaptive cellular genetic algorithm for mobility management of GSM-LTE cellular network users. Expert Systems with Applications, 106, pp. 290–304.

Dahi, Z.A., Alba, E. and Luque, G., 2022. A takeover time-driven adaptive evolutionary algorithm for mobile user tracking in pre-5G cellular networks. Applied Soft Computing, 116, p. 107992.

Gao, Y., Wu, Y., Hu, H., Chu, X. and Zhang, J., 2019. Licensed and unlicensed bands allocation for cellular users: A matching-based approach. IEEE Wireless Communications Letters, 8(3), pp. 969–972.

Krishnan, P., Jain, K., Poojara, S. R., Srirama, S. N., Pandey, T., & Buyya, R. 2024. eSIM and blockchain integrated secure zero-touch provisioning for autonomous cellular-IoTs in 5G networks. Computer Communications, 216, 324–345.

Lilhore, U. K., Dalal, S., & Simaiya, S. 2024. A cognitive security framework for detecting intrusions in IoT and 5G utilizing deep learning. Computers & Security, 136, 103560.

Maniak, T., Iqbal, R. and Doctor, F., 2018. Traffic modelling, visualisation and prediction for urban mobility management. In Advances in Hybridization of Intelligent Methods (pp. 57–70). Springer.

Mu, X., & Antwi-Afari, M. F. 2024. The applications of Internet of Things (IoT) in industrial management: a science mapping review. International Journal of Production Research, 62(5), 1928–1952.

Sapkale, P.V. and Kolekar, U.D., 2021. Modified naïve bayes classifier for mode switching and mobility management using cellular networks. Wireless Personal Communications, 116(3), pp. 2345–2366.

Shabbir, A., Rizvi, S., Alam, M. M., Shirazi, F., & Su'ud, M. M. 2024. Optimizing energy efficiency in heterogeneous networks: An integrated stochastic geometry approach with novel sleep mode strategies and QoS framework. PloS one, 19(2), e0296392.

Singh, S. and Prasad, D., 2021, September. Impact analysis of sink mobility on fault tolerance and delay in Wireless Body Area Networks (WBANs). In 2021 9th International Conference on Reliability, Infocom Technologies and Optimization (Trends and Future Directions) (ICRITO) (pp. 1–6). IEEE.

Sivasankar, S. and Challa, R., 2021, January. Closed loop paging optimization for efficient mobility management. In 2021 IEEE 18th Annual Consumer Communications & Networking Conference (CCNC) (pp. 1–7). IEEE.

Wang, C., Ma, L., Li, R., Durrani, T.S. and Zhang, H., 2019. Exploring trajectory prediction through machine learning methods. IEEE Access, 7, pp. 101441–101452.

Zhang, J., Xu, Y., Xia, W., Xu, Y., Cai, S. and Zhu, H., 2019. Edge caching and resource allocation scheme of downlink cloud radio access networks with fronthaul compression. IEEE Access, 7, pp. 118669–118678.

Chapter 10

Mobility Management through Machine Learning

*Nallakaruppan M K, Siva Rama Krishnan S,
Ramya G, Abdul Rehman Javed, Ishita Johri,
and Sweta Bhattacharya*

10.1 INTRODUCTION

The future world as envisioned today would have almost all devices virtually more intelligent making lives simplified and enriched. The role of artificial intelligence (AI) would play a significant role in such applications, which would emphasize delivering enhanced experiences and develop systems that would benefit our society better than ever. In comparison with third-generation mobile networking, 4G enabled enhanced quality video streaming but had associated challenges of network congestion.

The present generation, 5G, is completely data-driven with the evolution of smart cities and infrastructure management; multiple devices work simultaneously to ensure reliability, security, and uninterrupted services. Its benefits over 4G include higher speed, less latency, and capacity to connect multiple devices with less interference and enhanced efficiency. The new technology provides smart cities and intelligent transportation systems with a new method to being fully integrated by enabling numerous simultaneous connections and network ubiquity, especially in high mobility situations or densely populated regions. Several research projects, some funded by businesses and others by academic organizations, have looked at the integration of 5G technology with Internet of Things (IoT) services in smart city environments.

In contrast to current models, future 5G devices won't need a smartphone to access the Internet, making them fully connected. Already 5G is making homes smarter in terms of security (remote video monitoring and control and wirelessly controlled door locks), comfort, and convenience (voice control, remote control using smartphones, and thermostat management). The use of 5G for urban IoT may provide monitoring services for all city energy consumption, providing authorities with useful detailed information about the energy required for various public services (including, among others, street lighting, traffic lights, security cameras, and heating and cooling of public buildings). By doing so, it will be feasible to identify the main energy sources and develop plans to enhance city energy management.

DOI: 10.1201/9781003306290-10

133

Automobiles are also evolving into mobile sensor platforms that provide information to drivers about their surroundings that can be uploaded to the cloud. The sensor data are accessible to networks of autonomous vehicles that communicate with one another to maximize a clearly stated goal. In order to manage traffic more efficiently, with fewer delays, less pollution, and increased driver and passenger comfort, autonomous vehicles should cooperate if at all possible. The IoT, smart cities, and intelligent transportation systems are three of the major concepts that will be supported by the deployment of 5G technologies, especially in developing nations [1]. Figure 10.1 discusses the various components involved in 5G applications.

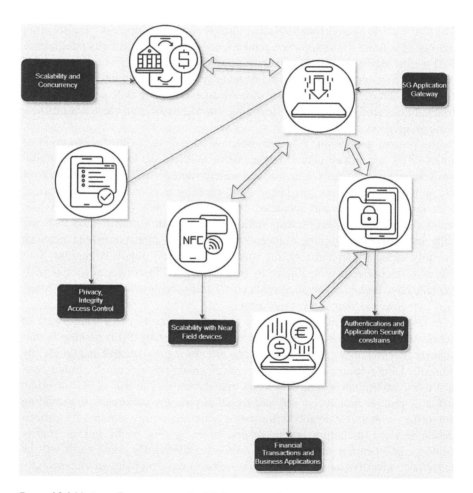

Figure 10.1 Various Components of a 5G Application Layer.

10.1.1 Challenges of 5G

With the advent of 5G in every domain, the major challenge remains the traffic management because of increased number of users and devices, but other challenges that need to be resolved pertain to peer-peer communication, dynamic change in the topology, and resource and compatibility constraints. Ultra-reliable low-latency communications (URLLC) applications require reliability for data delivery, privacy of data, and dedicated bandwidth with minimal delay. Retransmission mechanisms in URLLC can add to the overall delay while maintaining the reliability. The 5G network also uses enhanced mobile broadband (eMBB), which needs higher data transmission rates for data-intense applications. It would be challenging to maintain the requisite QoS if URLLC and eMBB were deployed in the same physical resource [2]. A small cell is a portable, miniature base station that is powered by very little electricity. To lessen propagation loss, thousands of these stations will be installed throughout smart cities. Figure 10.2 displays the

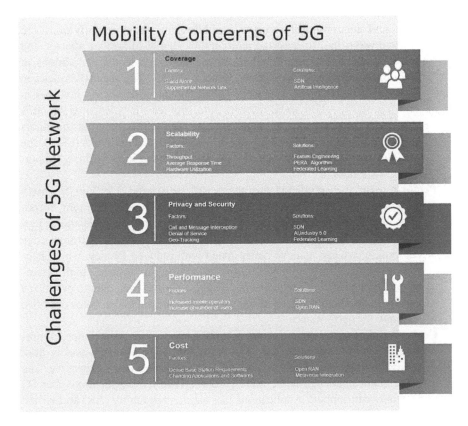

Figure 10.2 Challenges of 5G.

challenges related to implementing 5G today, such coverage, scalability, privacy, and security.

Coverage: The modern-era mobile devices are equipped with 5G coverage with advanced processors and radio device support. Companies like Qualcomm are the game changers in the industry, providing coverage in different bands such as standalone at 700Mhz and supplemental network link at 1400Mhz. The remaining challenges with 5G coverage are proximity-based connectivity, the short distances that signals travel compared with its predecessors, and reaching rural areas.

Scalability: This parameter is influenced by many factors such as throughput, processor utilization, average response time, number of devices requesting service in parallel, near-field communication, and channel interferences. Scalability in 5G is primarily focused on services like connection establishment and the user registration to the network. Concerns about a 5G network are mostly focused on its performance, but recent researchers on performance applied performance evaluation process algebra to try to address scalability [3].

Privacy and Security: The threat to a mobile network is generally produced in three ways, call or message interception, denial-of-service attacks, and geo-tracking. These threats remain unchanged even as the generations of mobile technologies evolved from 3G to 4G to 5G. The security issues can be handled in 5G using a concept called slicing. This slicing involves in splitting up the whole radio network into smaller slices, each of which is processed independently; an attack on one slice will not disrupt the others or the whole network. Concepts like software-defined networks (SDN), the IoT, and artificial intelligence (AI) can provide operational security support for 5G networks.

Performance: It is predicted that there will be around 2 billion 5G users worldwide by 2025. Therefore, there is a huge competition among corporates to grab the radio spectrum of 5G, but increasing channels affects performance because it increases the number of users per cell. Through edge computing, we can achieve better speed with reduced latency, which happens to enhance privacy and security in addition to assuring high performance. Open radio access networks (RANs) allow developers to produce a variety of applications in open source platforms, which also helps optimize 5G network performance.

Cost: Providing better coverage and optimal performance is the major requirement of any 5G network, and to achieve this, we need high-density base stations to provide optimal coverage, and these stations are expensive. The other major challenge is the more resource-intensive applications that 5G networks can support. Applications like augmented reality (AR) and virtual reality (VR) place high capacity demands on 5G networks that impose high system management and maintenance costs.

10.1.2 The Role of AI in 5G

AI has been predominantly used for solving complex problems that require hand-tuning when traditional approaches fail to provide appropriate solutions. AI and mobile and wireless systems are essential parts of social infrastructure development that help to mobilize our daily lives and also provides immense support for the digital economies. Although AI and 5G wireless communication are different fields of study, integrating the two technologies has huge potential to develop better applications, for instances in applications of supervised, unsupervised, and reinforcement learning.

In an Ericson poll of decision-makers from 132 cellular organizations globally, more than 50% of respondents expressed the opinion that combining AI with 5G technology would save capital costs, improve network performance, and create new revenue streams. The present generation of smartphones have become smaller, but the core algorithms have little improved since the 1990s. The 5G systems on the contrary consume excess power and deliver lower data rates, which do not coincide with the expected demand. The accelerated use of wireless technologies will thus eventually congest the airwaves with all the smart devices that are used to communicate with one another. As a result, the requirement for creating efficient communication devices that don't always broadcast on the same frequency arises.

As already known, 5G technologies work much faster than 4G by offering lower latency. The technologies also enable developers to create applications that work in enhanced response time, ensure seamless transmission of real-time video, and provide access to real-time data from various solutions. A subset of AI called machine learning (ML) develops algorithms and sophisticated statistical models that can carry out specified tasks without the use of human-like explicit instructions using patterns and conclusions. Without the application of AI and machine learning algorithms, a 5G network cannot operate fully or efficiently.

These 5G networks can be used to integrate AI and ML at the network edge. A large amount of data is produced by the simultaneous connection of several IoT devices thanks to 5G, which is then processed using ML and AI. AI and ML algorithms are also used to direct and manage application-based traffic across the access network. Additionally, with the integration of 5G with AI and ML, executing dynamic network slicing to address various situations with distinct QoS requirements is made possible. For flawless transmission utilizing 5G networks, the application of ML and AI enables 5G networks to become more proactive as well as predictive. Incorporating ML with 5G enables mobile devices to create adaptive clusters based on training data and gives intelligent base stations the capacity to make proactive decisions [4]. This increases the network applications' effectiveness, latency, and dependability.

10.1.3 Traffic Management in 5G Using AI

Due to novel scenarios, technologies, and network topologies, traffic management for 5G networks will provide significant technological challenges. The recent success of AI and ML technologies in a number of application domains has raised the possibility that they could help with the 5G traffic management problem. The novel aspects of 5G wireless network traffic management were covered in [5]. As prospective solutions and research directions for the management of 5G traffic, distributed and lightweight ML algorithms as well as a unique architecture for an AI helper content retrieval approach are examined.

It has been evident in recent years that for the next generation of cars, mobile networks will be essential for infotainment, traffic management, and autonomous driving applications. The next generation of intelligent vehicular networks made feasible by 5G systems will incorporate a number of heterogeneous wireless strategies in order to provide time-sensitive services with ensured quality of service and maximum bandwidth utilization.

The in-depth study of trustworthy and seamless wireless communication protocols in the context of vehicle networks is necessary, nonetheless, to enable the wide range of wireless technologies. Therefore, effective mobility management procedures that lessen the difficulties associated with the mobility of vehicles are crucial to supporting huge data loads across multiple applications. The authors of [6] examined several mobility management methods in 5G-enabled vehicle networks. The authors proposed categorizing different vehicular network models that are suitable for the 5G wireless network, followed by a full discussion of the mobility management problems that must be solved in each of these network models, before going over each one's benefits and drawbacks.

10.2 BACKGROUND

The Fourth Industrial Revolution will be a global, immersive experience thanks to 5G technology, which will also play a significant role in strengthening our global economy with a tech-based central nervous system. The technology will encourage the creation of automated capabilities for both tangible and intangible systems through the development of AI. As a result of a growing need for higher-quality mobile services in vertical industries, a range of technologies that have historically been alien to the telecom ecosystem, such as cloud computing, are being integrated into 5G [7]. One of the key and crucial components of 6G systems will be the use of AI and ML to futuristic wireless networks [8].

AI technology presents a different choice that may be better than conventional concepts and performance. Based on an understanding of the core 5G technologies, the authors of this paper [9] give an overview that first scans through a number of interesting research topics in AI for 5G technologies. They also provide design concepts including efficient resource allocation, unified 5G physical layer acceleration, and joint optimization of the entire physical layer.

The projected complexity of maintaining and managing 5G and beyond networks has driven the push toward closed-loop automation of network and service management operations. The ETSI Zero-touch network and Service Management (ZSM) framework [10] was created as a next-generation management system with the goal of totally automating all operational operations and responsibilities in order to accomplish this. AI is seen as a key facilitator of self-managing capacities, resulting in lower operational costs, a shorter time to value, and a lower chance of human error. Although there is a growing interest in using AI in ZSM systems, there are risks and limitations that should not be ignored.

In the face of ongoing issues with communication network architecture that are steadily becoming more complex and rising traffic demand over wireless systems, cognitive radio (CR) technology will still be insufficient for dynamic spectrum resource allocation in the 5G networks if it is intelligence-deficient. In order to create an efficient real-time process, this paper [11] proposes a distributed cognitive cellular network that fuses CR and AI into a sophisticated multiagent system. It is a novel paradigm for 5G wireless communication networks.

10.2.1 5G in IoT Applications

The IoT, smart cities, virtual reality, and many other applications are now supported in large part by the 5G network. In contrast to earlier network generations, 5G runs on higher radio frequencies and makes use of sophisticated digital technology like massive multiple-input/multiple-output antenna arrays. The introduction of new technologies and improved features in the 5G network has presented new challenges for network operators, and merging AI is one of the effective solutions to tackle these complications. However, a 5G network with AI capabilities presents security concerns and needs to be upgraded. Blockchain must be integrated to address these issues.

Blockchain can interact with other technologies to build smarter, more secure, and AI-based 5G cellular networks. In this article [12], Block5GIntell, a comprehensive intelligence and secure data analytics platform for 5G networks based on the convergence of Blockchain and AI, is introduced. The authors reach the conclusion that the Blockchain can combine with AI to develop smart and secure 5G networks using the provided framework.

The IoT is crucial to smart tourism. IoT applications for smart tourism are difficult to implement, though, because they need to handle massive amounts of data with low-latency connectivity. Researchers have described IoT systems for smart tourism that are 5G and AI enabled [13]. The authors use a point-of-interest (POI)-based reference system to analyze users' historical data to help build the POI attributes for next user.

Accelerated standardization of the 5G cellular network also means that more of the prospective technologies will be put into use. As a result, it is important to illuminate all feasible techniques and examine the guiding design ideas. Researchers [14] highlighted one of the core traits of the cutting-edge technologies employed in the 5G era, namely the emergence of initial intelligence in nearly all significant cellular network functions, such as radio resource, mobility, and service provisioning management.

Drones and assembly robots used in intelligent manufacturing are the main focus of recent robotics technology development. Along with the aforementioned application scenarios, the connection between AI and robots also contributes to the most important feature of intelligent transportation—unmanned driving. As 5G implementation is advancing, theoretical research [15] has begun on 6G. As mobile communication technology develops, the information transmission in the intelligent transportation system will become more dependable.

10.2.2   5G in Industrial Applications

Industrial production is evolving as a result of 5G and AI, and manufacturing businesses have a lot of opportunity. One of the effects of the expanding production data volume is the need for rapid data processing, fast transmission speeds, and the transfer of vast volumes of data. It is essential to combine 5G and AI technologies in order to flexible transmit enormous amounts of data and real-time data. In cooperation with the wbk (Institute of Production Science) of the Karlsruhe Institute of Technology in Karlsruhe, Germany, and Ruhr-University Bochum (Bochum, Germany), a concept [16] was developed at the Advanced Manufacturing Technology Center (AMTC) at Tongji University in Shanghai, China, and then put into practice in the AMTC learning factory.

Using the most recent developments in 5G and ultra-low-latency networking, as well as AI, robots, and other fields, the authors of [17] advocate for the formation of a completely new Internet that will enable the spread of talents in digital form. They outline the technical challenges that must be overcome to make this vision a reality, such as the development of a 5G tactile internet, standardized haptic codecs, and AI to support the perception of zero delay networks.

One area that is expected to drastically change with 5G services is e-commerce operations, particularly in combination with technologies like blockchain, AR, and VR. The faster speed of 5G networks and high-resolution screens of 5G-enabled devices could encourage customers to spend more time on e-commerce websites and potentially increase their online shopping. An enhanced customer experience and a vibrant e-commerce environment may arise from the combination [18][19].

10.2.3   Network Management in 5G

AI is anticipated to play a crucial role in enabling intelligent, adaptive, and autonomous security management in 5G and beyond networks because of its capacity to extract hidden patterns from a large collection of time-varying multidimensional data and deliver quicker and more accurate decisions. Unfortunately, AI is a two-edged sword that could also jeopardize the security of future networks due to its limitations [20].

This white paper [21] on AI/ML as enablers of 5G and B5G networks was developed with contributions from 5G public–private partnership initiatives that research, build, and validate 5G and B5G network technologies. Network diagnostics is focused on predicting network conditions, properties, and undesirable occurrences, such as security incidents. The method of estimating user location is one of the network insights. The white paper discusses how AI/ML will be used in the architecture of the 5G network. Solutions are identified for AI-based autonomous slice management, control, and orchestration; for scaling operations in network service orchestration; and other related problems.

Due to the rapid evolution of cellular system design toward 5G and beyond, it is necessary to examine the new features, design concepts, and solutions in realistic settings for various deployments and use cases. Although there are now a number of system-level simulators for 4G and 5G, a system-level holistic and realistic simulator that complies with 3GPP standards is urgently needed in order to evaluate the plethora of AI-based network automation solutions available. SyntheticNET is the first Python-based simulator [22] that fully complies with 3GPP 5G Standard Release 15 and can be upgraded to later editions. A crucial aspect of the platform is how easily AI-based network automation solutions can be evaluated using SyntheticNET. Because SyntheticNET is the first Python-based 5G emulator, it has the inherent ability to collect and analyze massive data sets and integrated access to ML frameworks, which contributes to its simplicity.

Rail communications have drawn much attention from the academic and industrial communities due to the rapid development of railroads, especially high-speed railroads (HSRs). In order to achieve the goal of future smart rail communications, the rail transportation industry must create cutting-edge communication network architectures and essential technologies that ensure high-quality transmissions for both passengers and railway operations and control systems. Fifth-generation technologies show promise in addressing the design issues for high reliability and high throughput for HSR communications thanks to their high mobility, security, comfort, and comfort, as well as their openness, predictability, and dependability. Researchers proposed a network slicing architecture for a 5G-based HSR system using a ray tracing-based analysis of radio wave propagation characteristics and channel models for millimeter wave bands in railroad conditions to obtain important findings about the appropriate operating frequency bands for HSRs [23].

According to the Japanese Ministry of Internal Affairs and Communications, the use of 5G for doctor-to-doctor remote diagnosis is progressing; patients can benefit from receiving state-of-the-art medical care without having to travel significant distances, and the expansion of doctors' service areas will make more specialists available. Deep learning-based AI is expected to assist in making this possibility a reality, which will also lighten the strain for medical practitioners. Diagnostic imaging using AI and an instruction-giving AI model will be able to provide real-time diagnostic support based on 5G [24]. This enables the use of remote diagnosis methods like endoscopy and ultrasonography, which do not collect information from fixed points like CT or MRI scans do. The AI model must provide instructions to the doctor visiting the patient at home in order to collect the relevant data in accordance with the patient's symptoms and condition.

Numerous new apps and services that will increase network traffic are targeted by the 5G system and beyond. Due to the aggressive, controversial, and competing requirements of these industrial verticals, it is difficult for the network to achieve its goals. It is anticipated that the requirements will include low temporal latency, large data rates, and reliable networks. Fortunately, telecom firms can have some hope thanks to the developments and advances in machine learning, cloud computing, micro-services, and the ETSI ZSM era. The development of a novel concept known as content-centric networking, which can adapt depending on the network and resource conditions as well as perceived key performance indicators, is a result of this, and business and academics are under enormous pressure to implement these techniques. [25].

Networks that are beyond fifth generation (B5G) are a recent innovation in communication technology. As a result of the extraordinary growth of AI, more intelligent B5G networks are projected to support the continual development of AI applications and the efficiency of communication technologies. More recently, B5G offshoots have caused billions of devices to send zillions of bytes of data to network edges. Training such massive data volumes using a centralized data center is a challenging task because of the extremely high bandwidth costs, low time efficiency, and considerable privacy leakages. It is also challenging to control and distribute the usage of learning results in a safe, dependable, and effective manner due to the variability and lack of confidence among edges. The solution that the professionals advise [26] for B5G networks is the AI-Chain, which unifies deep learning with blockchain and is powered by blockchain. New guidelines for sharing among edges can be developed using the AI-Chain, a distributed, immutable record of learning outcomes.

10.3   CASE STUDIES

This section highlights various applications of intelligent 5G in different real-time scenarios.

10.3.1   5G and AI in Manufacturing

Ericsson in association with Fraunhofer Institute of Production Technology [27] implemented 5G to improve the jet engineering manufacturing process, specifically to produce blade integrated discs (blisks). This type of production involves extremely demanding processes in which precision and accuracy play a major role. The process now in place is akin to a "black box," with no means of monitoring or correcting errors while production is still in progress and only the results of the finished milling process being made public.

The 3.5 GHz Ericsson 5G trial system is designed to address this issue when a vibration sensor is put on top of the blisk in the manufacturing equipment. Using 5G communication, the vibration spectrum is sent to the control system and assessed there in real time. Because of the short latency, it is feasible to correlate vibration with tool location and quickly alter the production process, resulting in faster control loops of less than 1 ms. The production of a digital twin—a virtual mirror of the component that will be manufactured and embodying the specifics of the process results—is a result of the continuous data transfer enabled by 5G. Such real-time monitoring involving ultra-low latency of less than a millisecond would ensure that any mechanical deviations are corrected immediately before it becomes critical acting as a major risk avoidance and mitigation activity.

The 5G simulation solutions developed by Altair [28] enables the development of wireless connectivity-based technology that improves communication, reduces latency, provides great coverage at reasonable cost, assures compatibility, and also reduces energy consumption and emission. One of the main products of Altair is a Car2X communication system based on vehicle-to-everything communication, which enables enhanced communication involving lower development costs. The simulation tools provided the necessary environment for comprehensive virtual test drive for 5G connectivity enabling autonomous driving and assessment of connectivity in real-time drive conditions.

Altair's 5G standard also acts as the foundation for various industrial IoT applications. The simulation of devices, 5G antennas, and 5G networks enables desired communication performance. The data rates for each application are achieved by identifying necessary changes and optimizing the antenna design and positions early, which avoids electromagnetic compatibility problems.

10.3.2   5G and AI in the Smart Grid

In Shenzhen, China, the introduction of 5G and artificial intelligence to the electrical grid radically changed all of the city's operations. A joint innovation lab for information and communication technology was created by the Shenzhen Power Supply Bureau (SPSB) and Huawei [29] with a primary focus on the study and use of cutting-edge technologies such as cloud, IoT, 5G, and AI in the electric power sector. A wholly owned subsidiary of China Southern Power Grid, which provides electricity to 2421 square kilometers, SPSB was founded in 1979.

Prior to the new technology developments, SPSB's work productivity was hindered by a lack of edge-device-based intelligent video identification capabilities. The original SPSB system's constrained capacity, depleted power supply, and depleted edge communication resources made it difficult to implement at a broad scale over the whole network. In order to create an intelligent inspection system for electricity transmission, SPSB collaborated with Huwaei. The new system implemented AI algorithms that allowed for the analysis of photos and videos, along with the Huwaei Atlas 200 AI acceleration component.

From their work, they generated alarms to help prevent potential dangers, and all transmission towers and frameworks, including drone monitoring equipment, have an AI processor powered by Ascent installed. Thus, the processors are able to recognize all five categories of possible dangers as well as seven significant pole and tower faults. This lowers the risks from potential onsite dangers and could result in extremely effective object detecting capabilities.

Another benefit of installing 5G intelligent inspection robots for substations is the ability to view data using high-definition cameras that are then relayed to big displays to set up an intelligent analysis system in the control center. Before 5G and AI, inspection took place only at night and was time-consuming and complex. Wired 4G sensor communication involved hardware communication facilities such gateways, optical distribution frames, auxiliary control devices, and electrical cables that required routine configuration and increased expenses. In order to monitor with greater speed, low latency, and bandwidth usage while delivering power device photos and infrared temperature data that permitted intelligent analysis in real-time, intelligent inspection robots with AI were used.

10.3.3   5G and AI in Mobile Applications

The year 2021 saw vast, impactful 5G interventions in the communication industry. Communication service providers (CSPs) upgraded operations to provide faster, more efficient, more flexible implementations by using the new 5G and other technologies and modernizing old systems, made possible by the proliferation of frequency bands and network densification. Demand increased for automation and digitalization of manual operations, which led to improved efficiency and expedited time to market.

One of the most well-known firms, Nokia [30], developed new processes and now integrates 5G and AI throughout its entire production and operations process from customer logistics to materials management, increasing transparency and lowering risks in its difficult adoption. When AI and ML are used for planning and resource management, project scheduling becomes incredibly flexible, resulting in exceptionally agile processes with improved quality. The Nokia platform's workflow has been digitalized, allowing for a smooth transition of drone-based digital site walk content and the creation of digital twins of the site, which results in automated site design. The use of AI and ML algorithms help in such development ensuring accelerated planning for future site expansions.

10.4   SECURITY IN 5G

This sections deals with the security concerns and defensive mechanisms in 5G using AI.

10.4.1   Security Issues

The capacity to preserve the confidentiality, integrity, and security of 5G is among the primary challenges with its design and deployment. Universal interconnectedness is tempting not just to the good actors who are supposed to utilize these networks but also to the bad actors that do use them. They have the potential to do damage from a distance that harms businesses, economies, and individuals as well as potentially national security. The security and privacy landscape has expanded beyond personal devices to service provider networks as a result of the larger bandwidth, greater spectrum usage, and high data rates of 5G networks.

In comparison with the networks that came before them, 5G networks support far higher levels of heterogeneity in terms of linked devices and networks. Smart cities, smart homes, and smart buildings are just a few of the smart technologies that are now possible with 5G networks, but this intensive interconnectedness increases the risks to an entire system of a security breach. Merged AI and ML techniques can spot weak security connections across various system components, including identity, authentication, and assurance and suggest solutions for enhancing the security and privacy of 5G IoT.

10.4.2   The Role of AI in Securing 5G Networks

Managing the safety and confidentiality of data coming from all of these different systems, each of which has its own specific security needs, is both challenging and onerous. AI and ML can support security mechanisms that provide comprehensive protection for the 5G IoT networks by examining the patterns and traits of network activity and learning to detect suspicious behaviors in real time. Security systems can monitor throughput and network error and use classification to identify and remove anomalies.

Clustering algorithms also make it possible to find and categorize weaknesses and threats to a network's security. Fictitious data sets using models like statistical inference attacks and generative adversarial networks (GANs) can be created and used to test, implement, and assess complex security protocols and measures. AI and ML can be used to handle the majority of programs, including those for antivirus scanner frameworks, interruption identification, spam channels, and misrepresentation detection frameworks; most of the time, the tactics work well with data from host processes, network traffic, and other sources. The service-based architecture that underpins the network for 5G and beyond, autonomous decentralized network services, and third-party servers pose threats in terms of DoS and cyberattacks. To better safeguard both the components and the system as a whole, specialized agents that are tailored to the domain of network components may be used.

One new threat is that recent research has shown that GANs can accurately mimic a network's output even when they are not provided access to the training data. In a single training process, the discriminator and generator DNNs compete with one another; the discriminator uses a game theory approach to decide if the fresh data samples produced by the generator are authentic or counterfeit. If a GAN is applied to mimic that network's output, the organization will never again be able to tell the difference between true and false information, which will allow the GAN to successfully create fresh examples of dissimulated information. In essence, without being exposed to the actual network, a GAN is capable of producing phishing data, deluging the core network with spam signals, and concocting a user authentication system. The continuous development of AI and ML for network security faces a number of issues since GAN has this potential to deceive the core network with proper authentication.

10.4.3   Defensive Mechanisms for Security Attacks in 5G

The various requirements of 5G technology can be realized with the assistance of algorithms that are based on artificial intelligence. This enables the technology to function in an automated and pre-programmed manner, which is necessary to meet the anticipated criteria for a higher quality of experience. The introduction of AI into 5G security helps with monitoring the traffic at SDN-enabled switches to track and identify a variety of fraudulent actions, such as man-in-the-middle attacks and radio jamming assaults. Tracking and understanding the past flow patterns including breaches and then analyzing current traffic allows for avoiding similar future assaults.

Techniques like spoofing and interception can attack wireless communications and transmission systems, but authentication is an essential component in shielding 5G-connected devices from threats. Traditional authentication techniques have generally relied on cryptographic processes, which have led to increased latency as well as higher computational expense. Furthermore, due to the amount of computing power that is needed for certain encryption techniques, these algorithms cannot be implemented on the devices. Fortunately, there are alternatives to traditional authentication methods.

Intelligent authentication that is built on top of AI gives the advantages of being situationally aware, having a very high dependability rate, and being cost-efficient all at the same time. The data from the accelerometer and the keystroke dynamics are used to construct a user's biometric identification via machine learning. This clone-resistant identification is independent not only of cell providers but also of the companies that make mobile devices. By merging both identities, it is possible to establish an encrypted authentication system, which prevents the controlling trusted authority from listening in on the user's private connections and prevents the user from being impersonated.

In light of the current state of designs for 5G wireless networks, we investigated many technical characteristics and services associated with 5G security, including availability, authentication, integrity, nonrepudiation, and confidentiality. AI holds both advantages and disadvantages for future networks' security, but developing new technologies and improving existing ones will assist us in overcoming this challenge.

10.5   AI APPLICATIONS FOR MOBILITY MANAGEMENT

This section deals with various AI techniques for mobility management and control.

10.5.1 Applications of 5G Bio-Inspired Computation

Bio-inspired algorithms in general provide solutions for feature engineering, optimization, and estimation of local and global best values for the evaluation of the fitness function for the target variable in a given problem. This field of AI is inspired from biological events such as ant colony optimization and moth-flame optimization (MFO). In this section, we discuss the possibilities of using bio-inspired computation to detect and select radio nodes. This process will be very useful for avoiding congestion and adjacent channel interference, increasing performance, and enhancing the QoS. This process detects communication nodes that are in active, idle, or power down mode and routes data to the nodes that are available to receive data, for instance with call forwarding.

Nodes that are not being used for communication can be withdrawn from the service, either switched off or put in power down mode, which improves the quality of the power management process and enhances the routing performance; MFO with simulated annealing (SA) is the process that supports this. MFO is an evolutionary computing algorithm that optimizes through minimization and maximization. It estimates the PBest, the best local solution, and a Gbest, the best global solution and the GBest by comparing it with the Pbest at the end of every iteration. However, because of the lack of exploration, the algorithm can give false-positive assumptions. Therefore we integrated SA with MFO [31] and developed a cascaded algorithm. This combination evaluates the fitness function, which is the GBest in lesser iterations, but increases the exploration with the help of the SA algorithm.

This nature of the MFO-SA combinations helps us in detecting mobile nodes that can actively participate in communication through the fitness evaluation. This can provide solutions for coverage, routing, traffic management, and cost optimization. Since the 5G networks require dense base stations, detection of unwanted nodes will enhance performance, mitigate congestion, and improves the QoS.

10.5.2 The Role of Explainable AI and Federated Learning in 5G Performance Enhancement

Packet tracing in a 5G network is a daunting task, since it is a dense and faster network. Thus, addressing data-related issues such as packet loss, duplication, and visual presentation of the packets is challenging in 5G. Enhanced AI can sift through results and present end users with exactly the data they want, which limits the unwanted data items users must deal with.

This process uses algorithms like RuleFit and tools like partial dependency plots to determine the relationship between two factors of analysis and decide how one factor impacts the other. For example, RuleFit can plot a graph between the packet loss and congestion and determine the linear relationship between the two. Thereby, we can visually understand the correlation between the two parameters and their impact on the target parameter. For packet tracing, global solutions can be provided by packages like Shapely.

This application allows us to view geometric objects without requiring any graphic packages. It always provides the centric view of the object even if the object is on a remote coordinate, which is extremely useful for packet tracing and finding the major causes of the packet loss. RuleFit can help eliminate unwanted packets that contain garbage data or corrupted packets by evaluating the feature-importance metrics for every packet.

Where explainable AI is helpful in packet-related applications, federated learning improves the privacy and security of 5G networks. Integrating federated learning with network data analysis [32] and multiparty communication protocol provides confidentiality, integrity, and privacy among mobile service providers, which is useful for gateway management at 5G mobile switching centers.

10.6 CONCLUSIONS

The emergence of 6G networks will definitely address the issues related to coverage and radio access to remote locations with increased bandwidth and capacity. However, when coverage and performance improve, challenges increase as well. The 6G network will provide higher capacity and less latency (1 ms) than 5G, its major feature, with its game-changing explainable AI with federated learning. This combination of tools solves issues related to routing, packet management, privacy, security, and cost. Thus, the next generation 6G networks will be able to provide increased scalability, performance, coverage, security, and QoS with the support of AI.

REFERENCES

[1] Guevara, L., & AuatCheein, F. (2020). The role of 5G technologies: Challenges in smart cities and intelligent transportation systems. *Sustainability, 12*(16), 6469.

[2] Siddiqi, M. A., Yu, H., &Joung, J. (2019). 5G ultra-reliable low-latency communication implementation challenges and operational issues with IoT devices. *Electronics, 8*(9), 981.

[3] Nallakaruppan, M. K., & Kumaran, U. S. (2020). Hybrid swarm intelligence for feature selection on IoT-based infrastructure. *International Journal of Cloud Computing, 9*(2–3), 216–231. doi: 10.1504/IJCC.2020.109375.

[4] www.deepsig.ai/how-artificial-intelligence-improves-5g-wireless-capabilities

[5] Fu, Y., Wang, S., Wang, C. X., Hong, X., & McLaughlin, S. (2018). Artificial intelligence to manage network traffic of 5G wireless networks. *IEEE Network, 32*(6), 58–64.

[6] Aljeri, N., & Boukerche, A. (2020). Mobility management in 5G-enabled vehicular networks: Models, protocols, and classification. *ACM Computing Surveys (CSUR), 53*(5), 1–35.

[7] Papagianni, C., Mangues-Bafalluy, J., Bermudez, P., Barmpounakis, S., De Vlee-schauwer, D., Brenes, J., . . . & Pepe, T. (2020, June). 5Growth: AI-driven 5G for automation in vertical industries. In *2020 European Conference on Networks and Communications (EuCNC)* (pp. 17–22). IEEE.

[8] Kaur, J., Khan, M. A., Iftikhar, M., Imran, M., & Haq, Q. E. U. (2021). Machine learning techniques for 5G and beyond. *IEEE Access, 9*, 23472–23488.

[9] You, X., Zhang, C., Tan, X., Jin, S., & Wu, H. (2019). AI for 5G: Research directions and paradigms. *Science China Information Sciences, 62*(2), 1–13.

[10] Benzaid, C., &Taleb, T. (2020). AI-driven zero touch network and service management in 5G and beyond: Challenges and research directions. *IEEE Network, 34*(2), 186–194.

[11] Wang, D., Song, B., Chen, D., & Du, X. (2019). Intelligent cognitive radio in 5G: AI-based hierarchical cognitive cellular networks. *IEEE Wireless Communications, 26*(3), 54–61.

[12] El Azzaoui, A., Singh, S. K., Pan, Y., & Park, J. H. (2020). Block5GIntell: Blockchain for AI-enabled 5G networks. *IEEE Access, 8*, 145918–145935.

[13] Wang, W., Kumar, N., Chen, J., Gong, Z., Kong, X., Wei, W., & Gao, H. (2020). Real-izing the potential of the internet of things for smart tourism with 5G and AI. *IEEE Network, 34*(6), 295–301.

[14] Li, R., Zhao, Z., Zhou, X., Ding, G., Chen, Y., Wang, Z., & Zhang, H. (2017). Intel-ligent 5G: When cellular networks meet artificial intelligence. *IEEE Wireless Com-munications, 24*(5), 175–183.

[15] Qiao, L., Li, Y., Chen, D., Serikawa, S., Guizani, M., & Lv, Z. (2021). A survey on 5G/6G, AI, and robotics. *Computers and Electrical Engineering, 95*, 107372.

[16] Zhang, W., Cai, W., Min, J., Fleischer, J., Ehrmann, C., Prinz, C., & Kreimeier, D. (2020). 5G and AI technology application in the AMTC learning factory. *Procedia Manufacturing, 45*, 66–71.

[17] Dohler, M., Mahmoodi, T., Lema, M. A., Condoluci, M., Sardis, F., Antonakoglou, K., & Aghvami, H. (2017, June). Internet of skills, where robotics meets AI, 5G and the Tactile Internet. In *2017 European Conference on Networks and Communications (EuCNC)* (pp. 1–5). IEEE.

[18] Kshetri, N. (2018). 5G in E-commerce activities. *IEEE IT Professional, 20*(4), 73–77.

[19] French, A., Shim, J. P., Risius, M., Larsen, K. R., & Jain, H. (2021). The 4th industrial revolution powered by the integration of AI, blockchain, and 5G. *Communications of the Association for Information Systems, 49*(1), 6.

[20] Benzaïd, C., & Taleb, T. (2020). AI for beyond 5G networks: A cyber-security defense or offense enabler? *IEEE Network, 34*(6), 140–147.

[21] Kaloxylos, A., Gavras, A., Camps, D., Ghoraishi, M., & Hrasnica, H. (2021). AI and ML–Enablers for beyond 5G Networks, AI/ML for Networks, 1–145. (https://www.recercat.cat/bitstream/handle/2072/522533/AI-MLforNetworks-v1-0.pdf?sequence=1)

[22] Zaidi, S. M. A., Manalastas, M., Farooq, H., & Imran, A. (2020). SyntheticNET: A 3GPP compliant simulator for AI enabled 5G and beyond. *IEEE Access, 8*, 82938–82950.

[23] Ai, B., Molisch, A. F., Rupp, M., & Zhong, Z. D. (2020). 5G key technologies for smart railways. *Proceedings of the IEEE, 108*(6), 856–893.

[24] Karako, K., Song, P., Chen, Y., & Tang, W. (2020). Realizing 5G-and AI-based doctor-to-doctor remote diagnosis: opportunities, challenges, and prospects. *BioSci-ence Trends, 14*(5), 314–317.

[25] Boudi, A., Bagaa, M., Pöyhönen, P., Taleb, T., & Flinck, H. (2021). AI-based resource management in beyond 5G cloud native environment. *IEEE Network, 35*(2), 128–135.

[26] Qiu, C., Yao, H., Wang, X., Zhang, N., Yu, F. R., & Niyato, D. (2020). AI-chain: Blockchain energized edge intelligence for beyond 5G networks. *IEEE Network*, *34*(6), 62–69.

[27] www.mmindia.co.in/article/868/5g-in-manufacturing

[28] www.altair.com/5g?utm_source=adwords&utm_term=5g%20manufacturing&utm_campaign=CO-5G-Search-AMER&utm_medium=ppc&hsa_mt=p&hsa_ad=553985474610&hsa_net=adwords&hsa_src=g&hsa_kw=5g%20manufacturing&hsa_tgt=kwd-842986999875&hsa_cam=14987278200&hsa_acc=8794474896&hsa_ver=3&hsa_grp=128601834477&gclid=Cj0KCQiApb2bBhDYARIsAChHC9vXNB1zd4Zu_7EQQvNxTg6S-HFLePq4MTWu5ZzVm6vuWD1-7LJ0-n0aAnoeEALw_wcB

[29] https://e.huawei.com/in/case-studies/industries/energy/2020/5g-ai-encounter-power-grids

[30] www.nokia.com/blog/5g-deployment-with-ai-ml-holds-big-promises/

[31] Arteaga, C. H. T., Ordoñez, A., & Rendon, O. M. C. (2020). Scalability and performance analysis in 5G core network slicing. *IEEE Access*, *8*, 142086–142100. doi: 10.1109/ACCESS.2020.3013597.

[32] Isaksson, M., & Norrman, K. (2020). Secure federated learning in 5G mobile networks. *GLOBECOM 2020–2020 IEEE Global Communications Conference*, pp. 1–6. doi: 10.1109/GLOBECOM42002.2020.9322479.

Chapter 11

Applying Heuristic Methods to the Offloading Problem in Edge Computing

Trong-Minh Hoang, Thu-Trang Ngo Thi, Hong-Hue
Nguyen Thi, Duc-Tan Tran, and Nam-Hoang Nguyen

11.1 INTRODUCTION

The emergence of network services and applications in 5G and beyond has increased the demand for extraordinary data transmission and processing capacity in computer and communication networks compared with earlier generations [1] [2]. These applications require mobile networks to meet high practical standards to achieve the performance requirements Internet of Everything, virtual reality, 3D applications, artificial intelligence, machine-to-machine communications, enhanced mobile broadband, and their supplementary technological directions. Most of these applications have stringent quality-of-service (QoS) requirements because of the explosive growth of network traffic. To address these issues, edge computing (EC) was created by transferring cloud-based computing and storage capabilities to the network's edge [3]. This reduces response time and enables users to offload computation duties with high bandwidth, low latency, and high quality of experience. Therefore, EC technology is essential for integrating IoT into 5G networks to surmount the previously mentioned challenges [4].

In an EC architecture, servers are frequently located in close proximity to end users, allowing them to make intelligent decisions based on their knowledge of local execution or cloud activities. This enables users to securely perform tasks such as computation and data processinga on their devices or through connected servers [5]. Notwithstanding the benefits of EC, substantial unloading difficulties must be addressed [6]. In order to achieve this purpose, researchers have investigated the offloading problem: that optimizing offloading is difficult because of the different tasks, multiple objectives, and environmental constraints of actual implementations.

Offloading has been identified as an NP-Hard problem [7, 8]. It has two primary objectives, minimizing energy consumption and minimizing response time, that conflict with one another: Reducing energy consumption delays response time and vice versa. Mathematical solvers, heuristic algorithms, Lyapunov optimization, game theory, the Markov decision process, and machine learning are methods for achieving these objectives. Implementing heuristic algorithms for large-scale problems may be inferior to mathematical solvers and machine

DOI: 10.1201/9781003306290-11

learning [9]. Consequently, the number of metaheuristics-based solutions has increased in recent years [10] [11].

There are many different heuristics algorithms used in a multiobjective, multiconstraint optimization problem. However, if intending to work at the network edge, the algorithm complexity is the highest priority to consider. Therefore, in this paper, we evaluate the performance of popular algorithms including Cuckoo, genetic algorithms (GAs), particle swarm optimization (PSO), teaching–learning-based optimization (TLBO), and Firefly in resolving the offloading optimization problem to find the most effective solutions. The contributions of this paper are twofold:

- Matching solutions are generated from heuristics algorithms involving the existence of actual constraints.
- The performance is evaluated of algorithms for meeting conflicting optimization goals, including energy and delay.

The structure of this paper is as follows: Next is a list of comparable works, and the third section provides a summary of heuristics approaches. Our proposed scenarios and validated outcomes are illustrated in Section 4. The paper concludes with a discussion of our prospective research and conclusions.

11.2   RELATED WORK

Due to the IoT's stochastically and constantly changing settings, optimizing offloading decision-making is complex, with several influencing factors and restrictions. Metaheuristic-based offloading optimization techniques have been developed to address the complexity and real-time requirements of realistic IoT systems. The primary purpose of task offloading is to reduce both the task execution time for end-user devices and the energy consumption of user equipment. These targets can be seen as a multiobjective, multiple-constraint generalization of the NP-hard problem.

Because of the dynamic and mobile nature of EC infrastructure, offloading is computationally exceedingly complicated, and the precise analysis of a complete solution space is impractical [12] [13] [14]. Consequently, heuristic approaches are frequently investigated because of their adequate running time for obtaining solutions. Heuristics provides sets of rules and procedures that assist in finding suitable solutions to computationally complicated situations. Their performance varies but has improved enormously in recent years [15] [16].

Delay limitation and power consumption reduction are the two most crucial aspects of the EC dispatching problem. Using the real-time constraint of services, the authors of [17] proposed a simulation annealing-based heuristic approach to minimize the task response time for the Internet of Vehicles. Using hybrid GA + PSO, the authors of [18] optimized the energy-focused service deployment strategy in an IoT environment. The authors of [19] proposed a hybrid heuristics method based on PSO to solve the problem in polynomial time, concentrating on task offloading and service caching to increase user satisfaction and resource efficiency. This proposal is a constrained discrete optimization problem.

Using a combination of a GA and the ant colony algorithm, the authors of [20] studied the optimization of multiaccess edge computing's (MEC) computational load from the perspective of delay and power savings. In [21], the authors found that a grayscale variation of the ant colony optimization algorithm, the whale optimization algorithm, and the wolf optimization algorithm was the best solution for offloading tasks utilizing well-known metadata. Various approaches based on GA have been utilized to simultaneously optimize the power and latency issues, as described in [22] and [23], with a somewhat detailed and realistic scenario. Through the abovementioned related research, we found that even though some hybrid heuristic algorithms can produce better results in certain circumstances, the complexity of the resulting solutions will be incompatible with the original objective. Edge computing is used for various practical applications, resulting in the efficacy of various offloading mechanisms.

To the best of our knowledge, there is no evaluation and performance comparison of the heuristic algorithms used to solve the offload problem after the survey. In this work, we will construct and verify heuristic algorithms for the offloading problem to evaluate the advantages and disadvantages of each strategy. This study examines the heuristic algorithms Cuckoo search [24], GA [25], PSO [26], Firefly algorithm [27], and TLBO [28]. We anticipate that the evaluation results, along with the no-free-lunch theorem [29], will generate fascinating research ideas in this field.

11.3 THE BACKGROUND

In this section, common metaheuristic algorithms for optimization are briefly introduced.

11.3.1 Cuckoo Search

Cuckoo search (CS) is one of the metaheuristic algorithms inspired by nature that was devised in 2009 by Shin-She Yang and Suash Deb [24]. Instead of straightforward isotropic random walks, the so-called Levy flight enhances the algorithm. Recent research indicates that CS has the potential to be significantly more effective than PSO, GAs, and other algorithms.

The solution to the CS algorithm is a population of nests or eggs. Each Cuckoo lays a single egg at a time and deposits it in a random nest. The nest containing the most eggs (solutions) will be passed to future generations. The number of accessible host nests is fixed, and a host bird's probability of detecting an alien egg is

Pa. In this scenario, the host bird can discard the egg or abandon the nest to start new eggs elsewhere. The CS pseudocode is listed in Algorithm 1 [30].

11.3.2 Genetic Algorithms

The most well-known evolutionary search algorithms are GAs. Initiated in 1975 by J. Holland to explain the adaptive processes of natural systems and design artificial systems based on them. GAs iteratively revise a population of individuals with a fitness function at each iteration. A new population is generated by selecting the best from the current generation. Some of these individuals are admitted unaltered into the following generation. Others are subjected to genetic operators such as crosses and mutation to produce new children. In the calculation, n represents the number of individuals in a population, the proportion of the population that will be supplanted by a crossover at each iteration, and m represents the mutation rate. There is a pseudocode illustration of a typical GA in algorithm 2 [31].

11.3.3 Particle Swarm Optimization

Kennedy and Eberhart proposed PSO in 1995 [26]. The concept underlying PSO is to use equations to simulate the swarming behavior of a collective group of entities, such as flying birds or a school of fish. Each entity adjusts its location as it moves at a certain speed and searches for food based on its intuition and the group's behavior.

In PSO, there are n particles or agents considered as a population. Particles move with specific velocities in the search space. For a particle or agent i, let x_i and v_i be the position (a.k.a. solution) and velocity, respectively. Each particle updates its position and velocity as:

$$v_i^{t+1} = v_i^t + \alpha\epsilon_1\left(g - x_i^t\right) + \beta\epsilon_2\left(x_i - x_i^t\right) \tag{1}$$

$$x_i^{t+1} = x_i^t + v_i^{t+1} \tag{2}$$

where ϵ_1, ϵ_2 are two uniformly distributed random numbers whose values are in [0,1]; α, β are learning parameters that are usually in the range of [0,2]; g is the best solution found so far by all particles in the population; x_i is the individually best solution during the entire past iteration history of the particle i. The pseudocode of the PSO is given in algorithm 3 [26].

11.3.4 Firefly Algorithm Optimization

Xin-She Yang devised the FA in late 2007 and published it in 2008 [27]. The algorithm was developed based on the biological behavior of tropical fireflies, which produce short, rhythmic flashes in distinct patterns to attract partners. By including both the inverse-square law of light intensity versus distance and the light absorption of the air, the FA becomes a nonlinear system whose descriptive equation is given as

$$x\_i^\wedge(t+1)=x\_i^\wedge t+\beta\_0\ e^\wedge(-\gamma r\_ij^\wedge 2)\ (x\_j^\wedge t-x\_i^\wedge t)+\alpha\epsilon\_i^\wedge t \tag{3}$$

where α is a scaling factor that controls the step sizes of the random movements; γ is a scale-dependent parameter controlling the visibility of the fireflies; the parameter β_0 is the attractiveness constant when the distance between two fireflies is zero. The pseudocode of the FA is listed in algorithm 4 [27].

11.3.5 Teaching–Learning-Based Optimization

In 2011 [28], Rao et al. proposed TLBO, a population-based algorithm. The activities in the educational classroom environment inspired this algorithm. The idea is based on the process by which the teacher guides the learners, and the learners share knowledge, which may facilitate a better outcome for a learner, i.e., the optimal solution.

The method includes the teaching phase and the learning phase. In the teaching phase, each individual x_i ($i = 1,2, \ldots, n$) (i.e., solution) learns from the teacher $x_{teacher}$, who is the best individual in the population. This can be understood as the learner moving their position toward the teacher and is described as

$$x_{new,i} = x_i + r\left(x_{teacher} - T_f x_{mean}\right) \tag{4}$$

where r is a random number in $[0,1]$; $T_f = round\left(1 + rand\left(0,1\right)\right)$ is a teaching factor; x_{mean} represents the average quality of all learners in the population.

In the learning phase, learners interact with one another to improve their performance through mutual knowledge sharing. For a random pair of learners x_i and x_j, x_i will move toward or away from x_j depending on whether x_j gives better performance than x_i does. Specifically, the new position of x_i while learning from its peers is given as

$$x_{new,i} = \begin{cases} x_i + r\left(x_j - x_i\right), & f\left(x_j\right) > f\left(x_j\right) \\ x_i + r\left(x_i - x_j\right), & f\left(x_j\right) < f\left(x_j\right) \end{cases} \tag{5}$$

The pseudocode of the FA is shown in algorithm 5 [28].

Algorithm 1 CS Optimization

- *Initialize the population by randomly generating n nests*
- *Evaluate the fitness value of each nest in the population based on the problem's objective function, which must be optimized.*
- *For each nest in the population:*

+ *Select a random parent nest from the population.*
  + *Generate a new offspring nest by applying a mutation or crossover operator on the parent nest, then evaluate the fitness value of the offspring nest.*
+ *If the fitness value of the offspring nest surpasses that of the parent nest, the offspring nest should substitute the parent nest in the population.*
- *For each nest in the population:*
  + *Create a new offspring nest by generating a random solution or modifying the current solution, and then assess the fitness value of the offspring nest.*
  + *If the fitness value of the offspring nest is better than the original nest, replace the original nest with the offspring nest in the population.*
- *Select the best nests, then update the new population to include only the best nests.*
- *Substitute some nests in the current population with the finest nests from the previous population to maintain the population's diversity.*
- *Continue the above steps until the termination criterion is met, such as reaching an optimal value or the maximum number of iterations.*

Algorithm 2 GA Optimization

- *Initialize the population by randomly generating n individuals*
- *Evaluate the fitness value of each individual in the population based on the objective function, which is the function that represents the problem to be optimized.*
- ***Repeat****:*
  + *Select individuals from the population for the next generation based on their fitness values based on a particular modality.*
  + *Create new offspring individuals by performing crossover between selected parents.*
  + *Perform mutation by randomly changing some chromosomes of individuals to promote the exploration of new solutions.*
  + *Evaluate the fitness value of the new offspring individuals and mutated individuals, then replace individuals in the current population with the new offspring and mutated individuals.*
Until: *the termination condition is met.*
- *Return the best individual as the optimized solution.*

Algorithm 3 PSO

- *Initialize the population by generating the particle positions and velocities at random.*
- *Set each particle's initial best position to its current position with the highest fitness value.*
- *Set the global best position as the position of the particle with the best fitness value.*

- **Repeat**: the
  + Evaluate the fitness value of each particle following the objective function, which represents the problem to be optimized.
  + If the fitness value of each particle's current position is more significant than its optimal position, update each particle's optimal position.
  + If each particle's best position has a higher fitness value than the global best position, update the global best position. Each particle's velocity is updated based on its current, best, and global best positions.
  + Each particle's position is updated based on its current position and velocity.

Until termination condition is met.
- Return the global best position as the optimized solution.

Algorithm 4 FA Optimization

- Initialize the population by randomly generating the positions of fireflies.
- Evaluate the attractiveness of each Firefly based on the objective function, which is the function that represents the problem to be optimized.
- **Repeat** :
  + For each Firefly i, calculate the attractiveness values concerning all other fireflies.
  + Update the position of Firefly i by moving towards fireflies with higher attractiveness values.
  + Update the location of Firefly i using equation (3).
  + Evaluate the attractiveness of each Firefly based on the updated positions.
  + Update the positions of fireflies based on the attractiveness values and movements.

Until termination condition is met.
- Return the best Firefly's position as the optimized solution.

Algorithm 5 TLBO

- Initialize the population by randomly generating n individuals
- Evaluate each individual's fitness value in the population based on the objective function, representing the optimization problem.
- **Repeat** :
  + Teacher phase:
    * Find the best individual (teacher) with the highest fitness value in the population.
    * For each individual in the population (except the teacher):
      • Generate a new solution by updating the individual's position using the teacher's position and other individuals' positions.
      • Assess the fitness value of the new solution.
      • Replace the individual's current position with the new solution if the fitness value of the new solution surpasses the individual's current fitness value.

+ *Leaner phase:*
  * *For each individual in the population:*
    • *Select a random individual (learner) from the population (excluding the current individual).*
    • *Generate a new solution by updating the individual's position using the learner's.*
    • *Evaluate the fitness value of the new solution.*
    • *Replace the individual's position with the new solution if the fitness value of the new solution is better than the individual's current fitness value.*
  + *Replace the worst individuals in the population with new randomly generated individuals to ensure diversity.*
Until. *the termination condition is met.*
- *Return the individual's best position as the optimized solution.*

11.4   THE PROPOSED OFFLOADING SCENARIOS AND NUMERICAL RESULTS

11.4.1   Assumptions

In order to demonstrate the efficiency of heuristic methods applied for offloading problems in MEC, a MEC model is simulated by Matlab. The general MEC model used in the simulation is illustrated in Figure 11.1. The set of end-user devices (UDs) is marked as $U = \{u_1, ..., u_i, ..., u_N\}$, which are distributed in the area

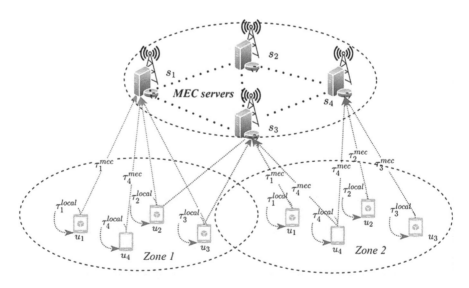

Figure 11.1 A General MEC Model.

with random distribution. MEC servers are donated as $S = \{s_1, ..., s_j, ..., s_K\}$. The zones (Zone 1, Zone 2, . . ., Zone Z) are defined according to base station-embedded MEC server coverage or clustering. The connection used in the model is a 5G radio link with its parameters and conditions followed by current technologies.

The task offloading process involves i) various hardware components, such as end-user devices and edge devices, ii) task splitting and computational processes, and iii) the transmission of data between the hardware components, which all can be executed locally and offloaded to a MEC server, called partial offloading at the edge.

We assume that each UD generates its tasks with the arrival rate λ_i according to the Poisson process and that one task must be offloaded in a single slot at a time. A set of computation tasks on UD_i is denoted as $\Gamma = \{\tau_1, \tau_1, ..., \tau_M\}, m \in M$. Each task, $\tau_i(m) \triangleq \langle l_i, e_i^{max}, t_i^{max} \rangle$, includes triple parameters, in which l_i is the length of the task (bits), e_i^{max} is the maximum energy and t_i^{max} is the deadline time. In this model, the end-user devices in the cluster zones of the 5G coverage areas access a MEC server based on the orthogonal frequency division multiple access, the transmission rate at which user device u_i transfers tasks to the server s_j is calculated as follows:

$$
r_{i,j}^u = \frac{W_{b_j}}{N} log_2 \left(1 + \frac{p_u \cdot h_{u_i,b_j}}{\omega_0 + \sum_{k=1, k \neq i}^{N} p_u \cdot h_{u_i,b_k}} \right) \tag{6}
$$

where W_{b_j} is the bandwidth of the server s_j linked to user devices, p_u is the transmission power of the user device u_i, h_{u_i,b_j} is the channel gain between the user device u_i and the server s_j, h_{u_i,b_k} is the channel gain between the user device u_i and its neighbor among N end-user devices, and ω_0 is the background noise.

Let's analyze the impact of a task offloading process on a task in the user device in the case of partial offloading at the edge. If task $\tau_i(m)$ is processed locally at the user's device u_i, the energy consumption and delay are computed by CPU cycle as follows:

$$
e_{i,u}^{local} = l_i^{local} \cdot C_u \cdot k \cdot f_u^2 \tag{7}
$$

$$
t_{i,u}^{local} = \frac{l_i^{local} \cdot C_u}{f_u} \tag{8}
$$

where $e_{i,u}^{local}$ is the energy consumption at u_i for the l_i^{local} bits, f_u is the CPU frequency, k is the capacitance constant, and C_u is the CPU cycle needed to process the task at the microprocessor in a user device.

Without loss of generality in case of partial offloading, p_i^{off} is denoted as the offloading proportion of the task τ_i, and p_s^{prob} is the probability of the MEC server executing the task offloading for a user device. The probability that a user device will choose a MEC server is dependent on the communication model and clustering strategy. Assuming that the capacity of the computation service of the MEC server is enough for multiple user devices, the executed energy and the executed time for offloaded tasks can be obtained as

$$e_{i,s}^{exe} = l_i^{mec}.C_s.k.f_s^2 = \left(p_i^{off} \times l_i \right).C_s.k.f_s^2 \tag{9}$$

$$t_{i,s}^{exe} = \frac{l_i^{mec}.C_s}{f_s} = \frac{\left(p_i^{off} \times l_i \right).C_s}{f_s} \tag{10}$$

where f_s is the CPU frequency at the MEC server, l_i^{mec} is the length of the offloading proportion of the task executed at the chosen MEC server, and C_s is the CPU cycle required to process the task at the MEC server.

Because the offloaded tasks need to be exchanged between UDs and MEC serves, the transmission time of a task τ_i is calculated as follows:

$$t_{i,j}^{trans} = \frac{l_i^{mec}.C_s}{r_{i,j}^u} = \frac{\left(p|i^{off} \times l_i \right).C_s}{r_{i,j}^u} \tag{11}$$

The energy consumption of transmitting data from UD u_i to MEC server s_j is formulated as

$$e_{i,j}^{mec} = p_s^{prob} \times p_i^{off} \times t_{i,j}^{trans} \tag{12}$$

Total energy required to perform a task τ_i in one session, including the local computational energy in a UD, the transmission energy, and the executed energy for offloading in a MEC server is

$$e_i^s = e_{i,u}^{local} + e_{i,j}^{mec} + e_{i,s}^{exe} \tag{13}$$

in which the total amount of energy consumed at one UD is in the form of

$$e_i^u = e_{i,u}^{local} + e_{i,j}^{mec} \tag{14}$$

The UD task is executed in parallel at the local device and the remote MEC server, and the executed latency of the task τ_i can be obtained as

$$t_i^u = max\left\{ t_{i,j}^{trans} + t_{i,s}^{exe}, t_{i,u}^{local} \right\} \tag{15}$$

At each UD, the resources for executing each task τ_i are constrained. Therefore, the ultimate optimization objective is finding offloading techniques that minimize latency and user energy consumption. Assume that the maximum tolerate time denoted as t_i^{max}, the maximum energy denoted as e_i^{max}, and the maximum transmission power denoted as p_u^{max} are represented for the upper bounds of resources for each task τ_i at a UD. Thus, the optimization function can be derived as follows:

$$maxC\left(p_i^{off}, p_u\right) = max \sum_{i=1}^{N} \left(\alpha E_i + (1-\alpha)T_i\right) \tag{16}$$

or can be detailed as

$$maxC\left(p_i^{off}, p_u\right) = max \sum_{i=1}^{N} \left(\alpha\left(\frac{e_i^{max} - e_i^u}{e_i^{max}}\right) + (1-\alpha)\left(\frac{t_i^{max} - t_i^u}{t_i^{max}}\right)\right) \tag{17}$$

subject to the following conditions

c1. $l_i = l_i^{local} + l_i^{mec}$

c2. $0 < p_i^{off} \leq 1$

c3. $0 < p_u \leq p_u^{max}$

c4. $t_i^u \leq t_i^{max}$

c5. $e_i^u \leq e_i^{max}$

c6. $0 < \alpha < 1$

The total cost, C, of an offloading method considers the balance between total energy consumption and latency. The weight, α, is the balancing factor of energy and delay. In different practical situations, the value of α will show that the obtained optimization goal of an offloading method greatly depends on saving energy consumption or delay time at the UD. The symbols and correspondent definitions used in the equations are illustrated in Table 11.1.

Table 11.1 Symbols and Definitions

Symbol	Description
α	Factor balancing energy and delay
λ_i	The arrival traffic rate of an end-user device
C_u, C_s	The CPU cycle at an end-user device and a MEC server
f_u, f_s	The CPU frequencies of an end-user device and a MEC server, respectively
l_i	The task's length τ_i
l_i^{local}	The length of bits processed by an end-user device
l_i^{mec}	The length of bits processed by a MEC server
$e_{i,u}^{local}$	The energy consumption for executing a task τ_i at an end-user device u_i

(Continued)

Table 11.1 (Continued)

Symbol	Description
$e_{i,j}^{mec}$	The energy consumption for transmitting a task τ_i from an end-user device u_i to a MEC server s_j
e_i^u	The total energy consumption at an end-user device u_i for a task τ_i
$e_{i,j}^{exe}$	The energy consumption for executing a task τ_i at a server s_j
e_i^{max}	The maximum energy of an end-user device u_i
e_i^s	The total amount of energy needed to perform a task τ_i in one session
t_i^u	The total latency to process a task τ_i
t_i^{trans}	The transmission time
t_i^{exe}	The time needed to execute at a MEC server
t_i^{local}	The time needed to execute at a local end-user device
t_i^{max}	The maximum delay at an end-user device u_i
p_u	The transmission power of an end-user device u_i
p_u^{max}	The maximum transmission power of an end-user device u_i
p_i^{off}	The proportion to offload a task τ_i
Γ	A set of end-user device tasks
S	A set of mobile edge computing servers
U	A set of end-user devices
N	Number of end-user devices in the system
K	Number of MEC servers in the system
M	Number of computation tasks on an end-user u_i

11.4.2 Numerical Validation and Comparisons

Matlab simulation validated the performances of the different heuristic EC offloading methods: Cuckoo single search and GA, FSO, Firefly, and TLBO population searches. Figure 11.2 shows the simulation scenario, and Table 11.2 summarizes the simulation's main parameters. Figure 11.3 shows that the cost performance of heuristic offloading algorithms varies with the iteration.

In general, the cost increases as the number of iterations increases and tends to remain constant at significant iterations in all cases. However, the performance of the single search method, Cuckoo, is always the worst of all algorithms at any iteration number. Among population search methods, PSO performs best, while GA performs worst. PSO costs increase sharply with fewer than 10 iterations but become saturated with more iterations. PSO cost more but achieved the fastest convergence. The performance curves of GA and Firefly are analogous; they gradually increase with the iterations. The cost of Firefly is always higher than the

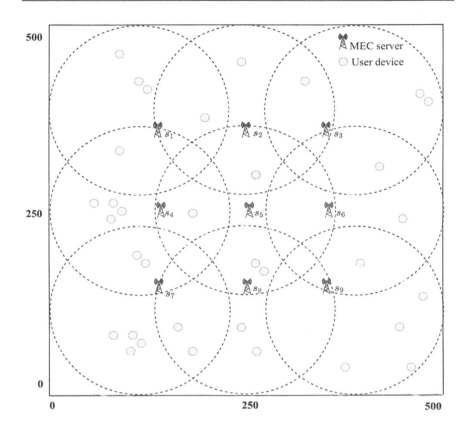

Figure 11.2 Simulation Scenario.

Table 11.2 Simulation Parameters.

Parameters	Value
The maximum iteration	50
Population size	100
Number of end-user devices	25
Maximum number of MEC Servers in range for a UD	3
Data volume of task (TaskSize)	40Kb; 100Kb ÷ 1000Kb
BS signal coverage	250
Edge server CPU frequency	10 GHz
BS bandwidth	10 MHz
User device transmission power	20 dBm
Background noise	−100 dBm

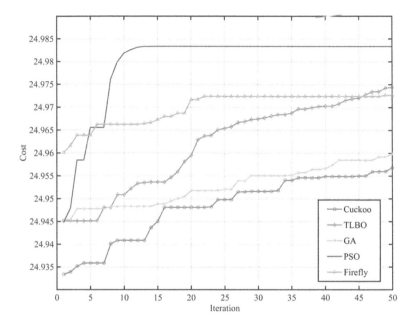

Figure 11.3 Cost vs. Iteration.

cost of GA with any iteration number. TLBO shows the most significant dependence of the total cost on the iteration number. With fewer than 20 iterations, the
cost of TLBO is close to that of GA; with more iterations, it approaches Firefly.

The weight, α, represents the contribution of the total energy consumption and
the latency to the cost performance of an offloading method. When α is small, the
latency is dominant in the cost. On the contrary, when α is high, the total energy
consumption dominates the cost. It can be seen from Figure 11.4 that these performances increase with the weight increase in all cases. This means the total energy
consumption affects the cost more strongly than the latency.

PSO has the highest cost at all weights. In addition, the contribution of the total
energy consumption and the latency to the cost in PSO is balanced. The performance of Cuckoo and GA are quite close to each other and are the worst of all,
and the change in their performance with weight is sharp: Cuckoo and GA benefit
at a very high cost if the total energy consumption prevails. The performance of
TLBO and Firefly is not as good as PSO's but better than Cuckoo's and GA's. All
methods investigated are distinguished at the small weights and tend to converge
at the high weights.

**The dependence of the consumption of a UD's total energy for one task on
the weight of heuristic offloading methods is demonstrated in Figure 11.5.**
The variation in the total energy consumption curves in terms of the weight is

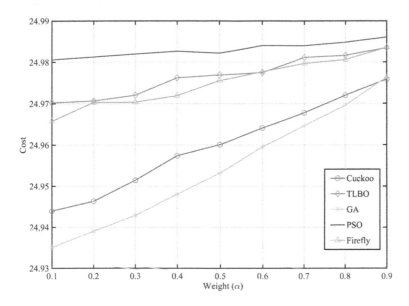

Figure 11.4 Cost vs. Weight (α).

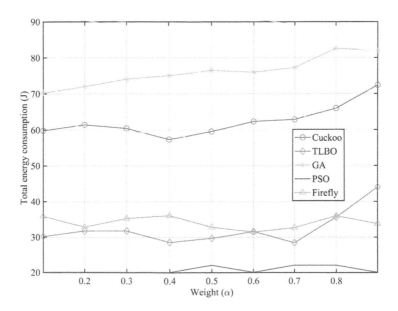

Figure 11.5 Total Energy Consumption vs. Weight (α).

similar in all cases. When α is small, there is a slight fluctuation in the total energy consumption because, in this weight region, the delay dominates the cost. In the high-weight region, the total energy consumption curves fluctuate dramatically.

The total energy consumptions of each offloading method are pretty distinguished. The total energy consumption of PSO is the smallest and changes little from the weight. The total energy consumption performances of TLBO and Firefly are quite close to each other and have a slight fluctuation. They are worse than PSO but much better than Cuckoo and GA. The total energy consumption of Cuckoo is much larger than that of PSO, and the total energy consumption of GA is the largest. Hence, total energy consumption strongly impacts the efficiency of heuristic offloading methods.

Figure 11.6 illustrates the dependence of total time consumption to process a task on the weight for all methods. With small weights, the total time consumption shows a bigger dependence on the weight, especially in the case of TLBO. With high weights, the total time consumption shows less dependence because, in this weight region, the total time consumption is not dominant over the cost. In the case of PSO and GA, the total time consumption is almost unchanged with the

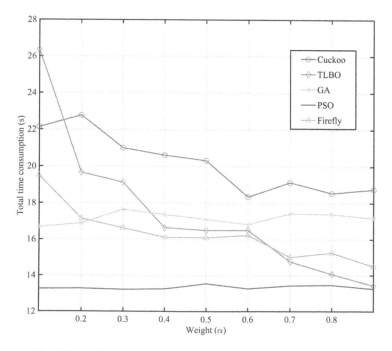

Figure 11.6 Total Time Consumption vs. Weight (α).

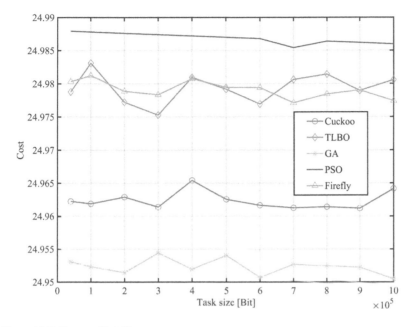

Figure 11.7 Cost vs. Task Size.

increase in weight. In addition, PSO shows the best behavior because its latency is always minimal with all weights.

On the contrary, the total time consumption of TLBO changes dramatically with the weight. At the weight region of small values, the performance of TLOB is worst. However, in the region of the high values, the performance of TLBO approaches that of PSO, the best one. The total time consumption of Cuckoo and Firefly decreases gradually with the weight, but the Cuckoo shows worse performance than Firefly. In general, there is a big difference in the latency of all methods at the small weights, but this difference tends to be narrower at the high weights.

It can be seen from Figure 11.7 that the efficiency of offloading algorithms on mobile edge computing with the task size. In general, the cost performances of all cases tend to decrease very slightly as the task size increases from 40 Kbit to 1 Mbit. This implies that the optimization goals of the heuristic offloading methods are still achieved at large task sizes. However, there are clear differences in the performances of these algorithms. With the same value of α, the cost of PSO is always the highest, while GA's is the lowest with all task sizes. The performances of TLBO and Firefly are very close to each other, not as good as that of PSO but much better than those of Cuckoo and GA.

11.5   CONCLUSION

With this paper, we evaluated the performance of low-complexity heuristic methods applicable to a partial load offloading model in a multiuser, multiserver mobile edge computing scenario. We also proposed a formulation for the dispatching problem's multiobject optimization. The simulation results indicate that PSO is the optimal solution for mobile edge computing outsourcing. In general, when evaluating the total cost, energy consumption, and task size for optimization, PSO, Firefly and TLBO, Cuckoo, and GA perform the best. However, the order of GA and Cuckoo changes regarding convergence and latency.

11.6   ACKNOWLEDGMENT

This work is the output of an ASEAN IVO project (https://www.nict.go.jp/en/asean_ivo/index.html) titled Agricultural IoT Based on Edge Computing and is financially supported by NICT (http://www.nict.go.jp/en/index.html).

REFERENCES

1. Shayea, Ibraheem, Mustafa Ergen, Marwan Hadri Azmi, Sultan Aldirmaz Çolak, Rosdiadee Nordin, and Yousef Ibrahim Daradkeh. "Key challenges, drivers and solutions for mobility management in 5G networks: A survey." IEEE access 8 (2020): 172534–172552.
2. Series, M. IMT Vision–Framework and overall objectives of the future development of IMT for 2020 and beyond. Recommendation ITU, (2015), 2083(0). https://scholar.google.com/scholar?hl=en&as_sdt=0%2C5&q=imt+vision+framework+and+overall+objectives+of+future+development&btnG=#d=gs_cit&t=1707390682647&u=%2Fscholar%3Fq%3Dinfo%3AT4zapDoRsnIJ%3Ascholar.google.com%2F%26output%3Dcite%26scirp%3D0%26hl%3Den
3. Yang, Ping, Yue Xiao, Ming Xiao, and Shaoqian Li. "6G wireless communications: Vision and potential techniques." IEEE network 33, no. 4 (2019): 70–75.
4. Liyanage, Madhusanka, et al. "Driving forces for Multi-Access Edge Computing (MEC) IoT integration in 5G." ICT Express 7, no. 2 (2021): 127–137.
5. Cao, Bin, et al. "Intelligent offloading in multi-access edge computing: A state-of-the-art review and framework." IEEE Communications Magazine 57, no. 3 (2019): 56–62.
6. Nguyen, Quang-Huy, and Falko Dressler. "A smartphone perspective on computation offloading—A survey." Computer Communications 159 (2020): 133–154.
7. Shakarami, Ali, Ali Shahidinejad, and Mostafa Ghobaei-Arani. "A review on the computation offloading approaches in mobile edge computing: A g ame-theoretic perspective." Software: Practice and Experience 50, no. 9 (2020): 1719–1759.
8. Jin, H., M. A. Gregory, and S. Li. "A review of intelligent computation offloading in multi-access edge computing." IEEE Access 10 (2022): 71481–71495. doi: 10.1109/ACCESS.2022.3187701.
9. Feng, Chuan, Pengchao Han, Xu Zhang, Bowen Yang, Yejun Liu, and Lei Guo. "Computation offloading in mobile edge computing networks: A survey." Journal of Network and Computer Applications (2022): 103366.

10. Jehangiri, Ali Imran, et al. "Mobility-aware computational offloading in mobile edge networks: A survey." Cluster Computing 24, no. 4 (2021): 2735–2756.
11. Sadatdiynov, Kuanishbay, Laizhong Cui, Lei Zhang, Joshua Zhexue Huang, Salman Salloum, and Mohammad Sultan Mahmud. "A review of optimization methods for computation offloading in edge computing networks." Digital Communications and Networks (2022), 9(2), pp. 450–461.
12. Lin, Hai, Sherali Zeadally, Zhihong Chen, Houda Labiod, and Lusheng Wang. "A survey on computation offloading modeling for edge computing." Journal of Network and Computer Applications 169 (2020): 102781.
13. Wang, Bo, Changhai Wang, Wanwei Huang, Ying Song, and Xiaoyun Qin. "A survey and taxonomy on task offloading for edge-cloud computing." IEEE Access 8 (2020): 186080–186101.
14. Islam, Akhirul, Arindam Debnath, Manojit Ghose, and Suchetana Chakraborty. "A survey on task offloading in multi-access edge computing." Journal of Systems Architecture 118 (2021): 102225.
15. Maray, Mohammed, and Junaid Shuja. "Computation offloading in mobile cloud computing and mobile edge computing: Survey, taxonomy, and open issues." Mobile Information Systems 2022 (2022).
16. Khan, Minhaj Ahmad. "A survey of computation offloading strategies for performance improvement of applications running on mobile devices." Journal of Network and Computer Applications 56 (2015): 28–40.
17. He, Yifan, Jing Xu, Bo Zheng, Jianqiang Hu, and Yong Xie. "Timing-oriented task offloading algorithms for internet-of-vehicles." Journal of Circuits, Systems and Computers 31, no. 8 (2022): 2250151.
18. Hu, Yuanchao, Tao Huang, Yang Yu, Yunzhu An, Meng Cheng, Wen Zhou, and Wentao Xian. "An energy-aware service placement strategy using hybrid meta-heuristic algorithm in iot environments." Cluster Computing (2022): 1-7.
19. Gao, T., Q. Tang, J. Li, Y. Zhang, Y. Li, and J. Zhang, "A particle swarm optimization with lévy flight for service caching and task offloading in edge-cloud computing." IEEE Access 10 (2022): 76636–76647. doi: 10.1109/ACCESS.2022.3192846.
20. Xu, Fei, Zengshi Qin, Linpeng Ning, and Zhuoya Zhang. "Research on computing offloading strategy based on genetic ant colony fusion algorithm." Simulation Modelling Practice and Theory 118 (2022): 102523.
21. Abbas, A., A. Raza, F. Aadil, and M. Maqsood. "Meta-heuristic-based offload task optimization in mobile edge computing." International Journal of Distributed Sensor Networks. https://doi.org/10.1177/15501477211023021.
22. Liao, Zhuofan, et al. "Adaptive offloading in mobile edge computing for extremely dense mobile networks based on genetic algorithms." Journal of Cloud Computing 10, no. 1 (2021): 1–16.
23. Wang, Jin, et al. "A priori probability offloading Priori mechanisms in mobile edge computing." IEEE Access 8 (2020): 39758–39767.
24. Yang, Xin-She. "Nature-inspired optimization algorithms: Challenges and open problems." Journal of Computational Science 46 (2020): 101104.
25. Hayes-Roth, Frederick. "Review of adaptation in natural and artificial systems by John H. Holland." The U. of Michigan Press, 1975. ACM SIGART Bulletin 53 (1975): 15–15.
26. Kennedy, James, and Russell Eberhart. "Particle swarm optimization." Proceedings of ICNN'95-International Conference on Neural Networks 4 (1995): 1942–1948. IEEE.

27. Yang, Xin-She. Engineering Optimization: An Introduction with Metaheuristic Applications. John Wiley & Sons, 2010. Oxford, England OX4 2DQ.

28. Rao, R. Venkata, Vimal J. Savsani, and D. P. Vakharia. "Teaching–learning-based optimization: a novel method for constrained mechanical design optimization problems." Computer-aided design 43, no. 3 (2011): 303–315.

29. Wolpert, David H., and William G. Macready. "No free lunch theorems for optimization." IEEE Transactions on evolutionary computation 1, no. 1 (1997): 67–82.

30. Lim, Wei Chen Esmonde, Ganesan Kanagaraj, and S. G. Ponnambalam. "Cuckoo search algorithm for optimization of sequence in pcb holes drilling process." In Emerging Trends in Science, Engineering and Technology: Proceedings of International Conference, INCOSET 2012, pp. 207–216. Springer India, 2012.

31. Jebari, Khalid, and Mohammed Madiafi. "Selection methods for genetic algorithms." International Journal of Emerging Sciences 3, no. 4 (2013): 333–344.

Chapter 12

AR/VR Data Prediction and a Slicing Model for 5G Edge Computing

Vithya Ganesan, Viriyala Sri Anima Padmini, V. Anjana Devi, Subrata Chowdhury, Gautam Srivastava, and Kassian T.T. Amesho

12.1 INTRODUCTION

Edge computing is a distributed computing paradigm that brings computation and data storage closer to the devices or users generating and consuming the data. This is particularly relevant in the context of 5G networks, which support a wide range of applications with diverse requirements in terms of latency, bandwidth, and reliability.

One potential application of edge computing in 5G networks is augmented reality (AR) and virtual reality (VR). These types of applications often require real-time processing and low latency to provide a seamless and immersive experience to the user. By bringing computation and data storage closer to the user, edge computing can help to reduce the round-trip time between the device and the server, improving the overall performance and user experience.

One way to achieve this is by data prediction and slicing models. Data prediction involves using machine learning algorithms to anticipate the data that a device will need based on its current context and past behavior. This can help to reduce the amount of data that needs to be transmitted over the network, improving efficiency and reducing latency. Slicing, on the other hand, involves partitioning the network into smaller, independent units or "slices" that can be customized to meet the specific requirements of different applications or user groups. This can help to ensure that the resources of the network are used efficiently and that the quality of service (QoS) for different applications is optimized.

Overall, edge computing and data prediction and slicing models have the potential to significantly improve the performance and user experience of mobile AR and VR applications in 5G networks

12.1.1 Literature Survey

Applications such as VR and AR require high bandwidth and low latency. Data prediction and slicing [1] Methodology improves edge computation by resource allocation, reduce offloading workload. Machine learning algorithms is required to optimize the performance of mobile AR and VR services in 5G

DOI: 10.1201/9781003306290-12

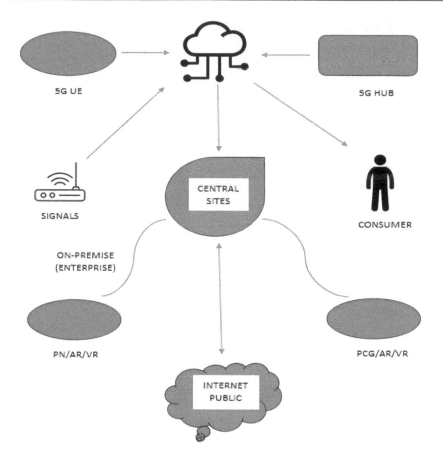

Figure 12.1 5G Edge Computing Architecture.

networks [2]. Slicing model may be helpful migration, offloading, resource allocation, content caching, mobility, computing and mobility model, system model, virtualization [3].

Adaptive learning framework which slicing the architecture for resource management and load prediction in data driven beyond 5G is developed for efficient slicing [4]. A framework is developed with three main blocks such as: Collector, Processor, and Decision Maker are used for data slicing methodology. The Collector module which initiates data acquisition and pre-processing by the components [5] as

- Data Collection: which collect the data of QoS metrics such as Latency, delay, Jitter, Packet Loss, Reliability, Bandwidth.

- Data Cleaning: Cleans the data, the fixing accessing errors, and structuring of dataset.
- Validation: verification of data expected from a dataset.
- Transforming: It is three subcomponents such as Scaling, Typing, and Encoding.

Furthermore, Edge computing is Integrated with 5 G by data slicing. For example, IoT and delay sensitive IoT demands edge computation to transfer delay-insensitive data to improve edge computation, reduced communication cost, and increased scalability. Hence network slicing is needed in edge computation to support faster security features in its proximity to the end devices and it is used in autonomous driving, seamless Internet, and industrial Internet, demand ultra-low latency, high reliability, and prioritization. [6].

Due to some inherent limitations of edge computing for AR/VR data storage, bandwidth requirement for offloading, CPU time [7]and for resource management a framework is required to deploy smart applications. Improved temporal and spatial characteristics will support AR/VR edge data computations [8]. In this consequence, edge computing divides the tasks to maintain the functional integrity, each part segmentation between the chunk of AR/VR data to facilitate resource allocation by code parser, system parser, and decision engine with AR/VR data slicing. First, code parser determines the application and code/data partition. Second, the system parser for monitoring available bandwidth, maximum size of data to offload, or the energy consumed to execute the local application. Third, the decision engine determines the status of offload as local execution, full offloading, and partial offloading [9] to support low latency, mobility, geographical distribution, strong wireless connectivity, and a high number of nodes [10].

To improve the offload/rendering at the edge to streaming the video and audio to increase the end-to-end delay for offloading augmented reality (AR)/virtual reality (VR) by prioritizing data at the level of a sub-flow and scheduling for reducing latency [11]. It is necessary to integrate edge Computing, 5G network slicing, High-Definition Maps and high accuracy positioning to provide fast offloading, decision-making and control parameters by using machine learning is required [12] in addition with a framework for deep learning-based approach to predicting mobile AR and VR traffic in 5G edge computing networks for resource allocation and network slicing.

12.1.2 Methodology

Edge computing in 5G networks support mobile augmented reality (AR) and virtual reality (VR) applications to evaluate the effectiveness of data prediction and slicing models by three phenomena. They are Collector, Scheduler, Decision controller.

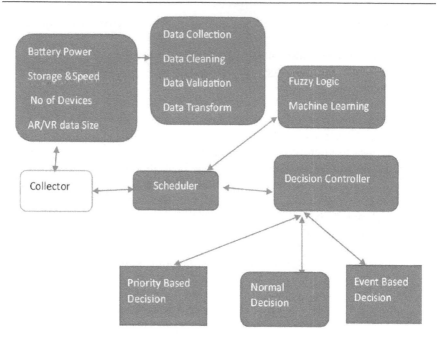

Figure 12.2 Data Slicing and Prediction by Machine Learning.

12.1.2.1 Collector for Offloading

Edge computing collector parameters are Battery Power, Storage and Speed, no. of devices, AR/VR data Size, and Bandwidth. Collectors require Data Collection, Cleaning, validation, and transforming for Edge computing.

$$Ec= \sum Bp + Sc+ Nn+ AR/VR \; sz+ B_{Avail}$$

Where Bp= Battery Power
Sc= Storage capacity
AR/VR = Data Size
$B_{Avail} =$ Bandwidth Available

12.1.2.2 ML for Scheduling

Schedulerdivide the resource types into IO type, intermediate speed and CPU type, and time span and load balance. These are the indicators to complete the edge computing scheduling. Calculate the computing power of each resource node weightfor AR/VR data by

$$Sc= w_{1\,x} n(Io \; Type) + w_{2\,x} p(CPU \; power)+ w_{3\,x} T \; (time) + w_{4\,x} l(load \; capacity)$$

Where w1, w2, w3, w4 are weightages for the parameters

$N, p, T, l = $ *quantity or unit of parameters*

Scheduler schedules the AR/VR data slicing by Fuzzy Logic and Machine Learning techniques algorithm

12.1.2.3   ML Resource Allocation

Decision Controller: AR/VR data slicing is operated on 3 modes namely priority-based, normal, and event-based decision controller. Taking the occupied space, load fairness, Allocation rationality, resource balance, the longest queue length and energy consumption value are taken as mode to assign the priority level. From the data set, parameter's level and its status are tabulated below:-

L1- Low, L2- Medium, L3–High

Table 12.1: shows the data collection and cleaning model for data slicing and prediction model.

12.1.3   Implementation

In Figure 12.3, we present a visual overview of the data employed in the context of 5G data slicing and prediction. This visualization provides a comprehensive understanding of the dataset, emphasizing several key aspects:

1. Normalized Overview of Data: The figure showcases a normalized representation of the dataset, ensuring that the data is transformed into a common scale for consistency and ease of analysis. Normalization is an essential step in data preprocessing, enabling meaningful comparisons and insights.
2. 16 Predictors Classified: The figure highlights that there are 16 predictors within the dataset. Predictors are the variables or attributes that are used to make predictions or classifications, and they play a crucial role in the data slicing and prediction processes.
3. Classifying Variables in Data Set: The figure also categorizes and classifies the variables present in the dataset. This classification is essential for understanding the role and significance of each variable in the prediction and slicing models.

Table 12.1 Modes for Slicing

Parameters for AR/VR data Slicing	L1	L2	L3	Mode for Slicing
More Occupied space for AR/VR data			√	priority-based
Load fairness	√			Normal
Allocation rationality		√		Normal
Resource balance			√	priority-based or event-based
Longest queue length	√			priority-based
Energy consumption		√		priority-based or event-based

Overall, Figure 12.3 serves as an informative visual guide to the dataset used in 5G data slicing and prediction. It emphasizes the importance of data normalization and the presence of 16 predictors, setting the stage for further analysis and modeling in the context of 5G network optimization.

The range sequence heat map provides a visual representation of the dataset after cleaning and validation. This heatmap allows us to visualize the distribution and range of data points. It helps in identifying any remaining anomalies or patterns within the data that may impact the slicing and prediction models.

The clean and validated dataset is a critical input for the machine learning-based algorithms that power AR/VR data prediction and slicing models, which are essential components in optimizing 5G network utilization for AR/VR experiences. These models are designed to anticipate data requirements, reduce latency, and enhance the overall user experience. Figure 12.4 serves as the starting

```
###############################################################################
Classifying variables in data set...
Data cleaning improvement suggestions. Complete them before proceeding to ML modeling.
```

	Nuniques	dtype	Nulls	Nullpercent	NuniquePercent	Value counts Min	Data cleaning improvement suggestions
Time	24	int8	0	0.000000	0.075990	0	
LTE5gCategory	22	int8	0	0.000000	0.069658	0	
Packetdelay	7	int16	0	0.000000	0.022164	0	
PacketLossRate	3	float16	0	0.000000	0.009499	0	skewed: cap or drop outliers
IoT	2	int8	0	0.000000	0.006333	0	
LTE5G	2	int8	0	0.000000	0.006333	0	
GBR	2	int8	0	0.000000	0.006333	0	
NonGBR	2	int8	0	0.000000	0.006333	0	
ARVRGaming	2	int8	0	0.000000	0.006333	0	
Healthcare	2	int8	0	0.000000	0.006333	0	
Industry40	2	int8	0	0.000000	0.006333	0	
IoTDevices	2	int8	0	0.000000	0.006333	0	
PublicSafety	2	int8	0	0.000000	0.006333	0	
SmartCityHome	2	int8	0	0.000000	0.006333	0	
SmartTransportation	2	int8	0	0.000000	0.006333	0	
Smartphone	2	int8	0	0.000000	0.006333	0	

```
16 Predictors classified...
   No variables removed since no ID or low-information variables found in data set
```

Figure 12.3 Normalized Overview of the Data.

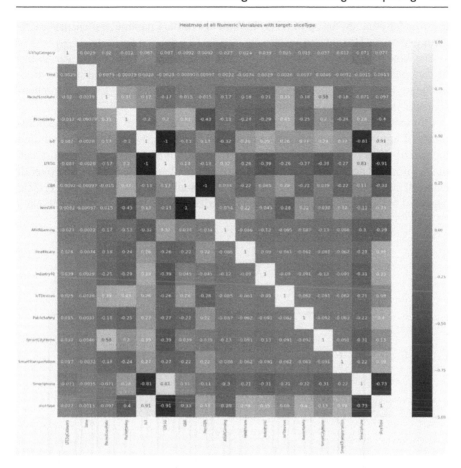

Figure 12.4 Data Cleaning, Validation, and Range Sequence (Heat Map).

point in the journey towards efficient AR/VR data delivery in edge computing environments.

The heat map utilizes color gradients to indicate the strength and direction of correlations. Darker shades represent stronger positive correlations, while lighter shades signify weaker or negative correlations. By analyzing this heat map, we can gain insights into which numeric variables have a significant impact on the choice of slice type.

Understanding the relationships between numeric variables and the target slice type is crucial for optimizing AR/VR data slicing strategies. This information can inform decisions on how to allocate network resources effectively, ensuring that the chosen slice type aligns with the specific requirements of AR/VR applications.

In Figure 12.5, we present a box plot that illustrates the scheduling of data parameters within the 5G spectrum. This visualization is a critical component of the broader strategy for optimizing the transfer of data parameters, ensuring efficient use of the 5G network resources.

The box plot provides valuable insights into the distribution and variability of data parameters being transferred. It depicts important statistical information, including the median, quartiles, and potential outliers, allowing for a comprehensive analysis of the data's characteristics.

Efficient scheduling of data parameters is essential to meet the diverse requirements of applications and services running on the 5G network. By understanding the distribution of these parameters, network administrators and operators can make informed decisions about resource allocation, latency reduction, and overall network performance. Figure 12.5 aids in the visual representation of this critical aspect of 5G network management and optimization. In Figure 12.6, we present a

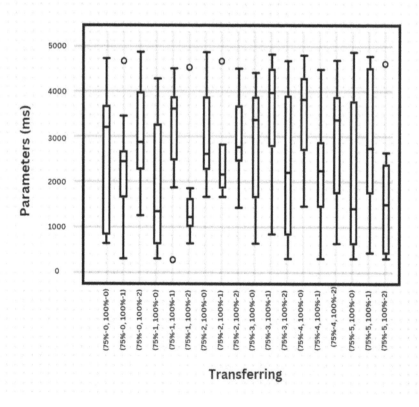

Figure 12.5 Scheduler for Data Parameters Being Transferred (Box Plot).

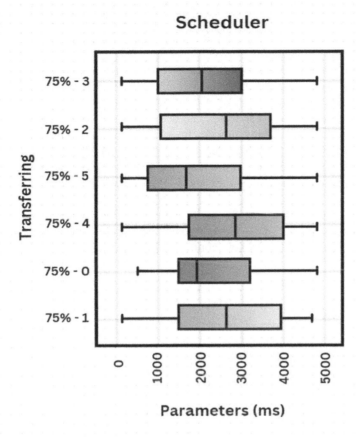

Figure 12.6 Scheduler (75%–100%) Data Transfer (Box Plot).

box plot that showcases the scheduling dynamics between specific parameters and the transfer of data parameters within the 5G spectrum. This visualization offers a comprehensive view of how these parameters are managed and optimized to ensure efficient data transfer in the 5G network.

The box plot in Figure 12.6 represents the relationship between data parameters and the scheduler. The y-axis, labeled "Scheduler (75%-100%)," displays the scheduling aspect, while the x-axis represents the data parameters being transferred. The box plot provides valuable statistical insights, including the median, quartiles, and potential outliers, allowing for a detailed analysis of how different data parameters are scheduled and transferred.

Efficient scheduling between parameters and data transfer is a critical aspect of 5G network management. This visualization assists in understanding how data parameters are managed and allocated within the network, ensuring that data is delivered with the required quality of service. Figure 12.6 helps network administrators and operators make informed decisions to optimize 5G network performance and resource utilization.

In Figure 12.7, we illustrate the transfer of control and decision-making processes between two key variables: "Petal Width" and "Grid Control." This visualization represents a decision controller using a decision tree to make decisions based on the values of these variables. The decision tree visually depicts how the control and decision-making process flows based on the values of "Petal Width" and "Grid Control." Each branch and node in the decision tree represents a specific condition or decision point. The tree structure helps determine the sequence of decisions and actions to be taken, depending on the input values of these variables.

This figure is a valuable tool for understanding the decision-making logic within a specific context, where "Petal Width" and "Grid Control" are crucial factors. It provides insights into how control and decisions are interrelated and

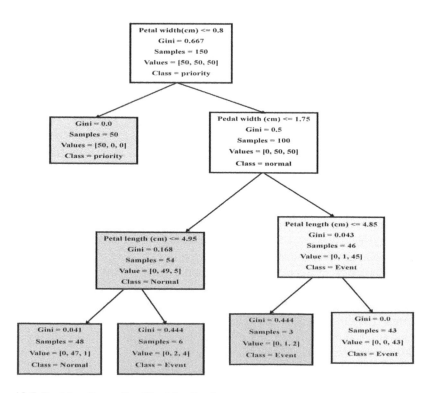

Figure 12.7 Decision Controller (Decision Tree).

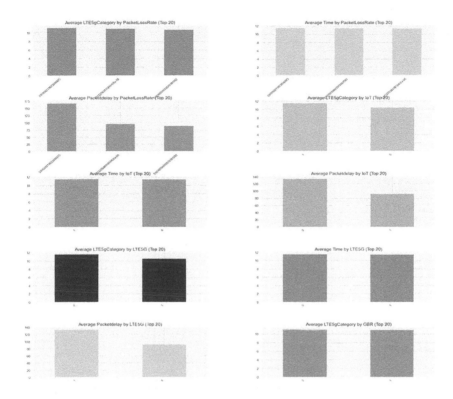

Figure 12.8 Data Visualization of Network (5G).

influenced by these variables, making it easier to interpret and optimize decision-making processes in this scenario.

Figure 12.8 contains multiple graphs or charts that display the "Average Packet Loss Rate" across various parameters. These charts provide a visual comparison of how packet loss rates vary under different conditions or scenarios. Each chart likely focuses on a specific parameter or set of parameters, allowing for a more detailed analysis of packet loss within the 5G network. This level of granularity in the data visualization can be instrumental in identifying trends, patterns, and potential areas for improvement in network performance.

In Figure 12.9, we present a histogram analysis that focuses on the distribution of multivariate variables. This visualization provides insights into the distribution and characteristics of data within a multivariate context. Each histogram represents the frequency and distribution of specific variables, allowing for a comprehensive view of how these variables interact and contribute to the overall dataset. The use of histograms in multivariate analysis enables a more in-depth exploration of data patterns, relationships, and dependencies among multiple variables. This comprehensive analysis aids in understanding how various parameters influence

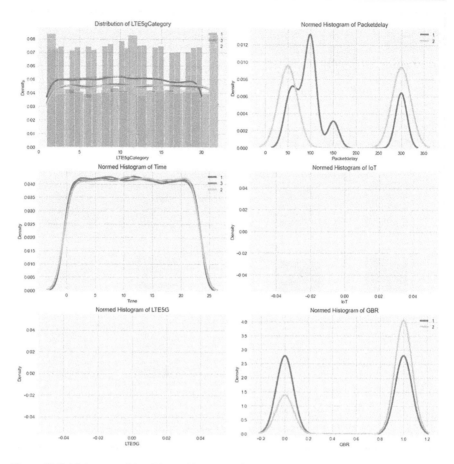

Figure 12.9 Multivariate Variables in Histogram Analysis.

the distribution of the dataset, providing valuable insights for data analysis and decision-making processes.

In Figure 12.10, we depict the network of cloud data content. This visualization illustrates the interconnections and relationships between various elements of cloud data content, highlighting how data is organized and distributed within a cloud-based environment.

The network diagram provides a visual representation of the structure and flow of data within the cloud, showcasing how different components interact and contribute to the overall ecosystem. This understanding is essential for effective cloud data management and optimization. Figure 12.10 aids in visualizing the complexity of cloud data content networks and their implications for data storage, retrieval, and access.

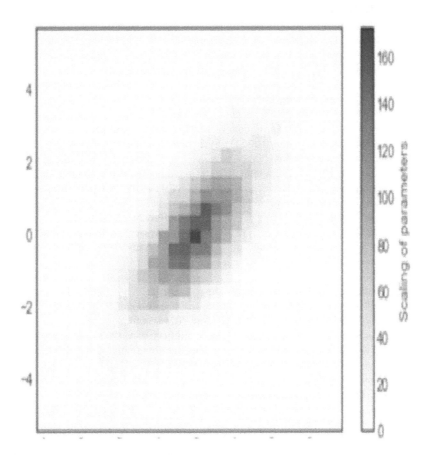

Figure 12.10 Network Cloud of Data Content.

12.2   CONCLUSION

To shorten the edge computing AR/VR data by collector, scheduling, and resource allocation by improve the load balance, resource scheduling method based on improved ML is proposed. Calculate the maximum communication rate for resource scheduling model by considering the size, type, and distribution of data to ensure the relative integrity with resource entity relationship. In the form of adaptive ML prediction, the sample scheduling feature association pair is enabled to extract the AR/VR slicing ratio. It expands the scale of edge computing to improves the scheduling ability of the whole platform, makes full use of resources. In the future research work, we should further strengthen the research on the efficiency of resource scheduling, and comprehensively improve the performance of edge computation methods.

REFERENCES

1. Mourtzis, D., Angelopoulos, J., & Panopoulos, N. (2021). Smart manufacturing and tactile internet based on 5G in industry 4.0: Challenges, applications and new trends, Electronics, 10, 3175. https://doi.org/10.3390/electronics10243175.
2. Haidine, A., Salmam, F. Z., Aqqal, A., & Dahbi, A. (2021). Artificial intelligence and machine learning in 5G and beyond: A survey and perspectives. In Moving Broadband Mobile Communications Forward—Intelligent Technologies for 5G and Beyond. IntechOpen. https://doi.org/10.5772/intechopen.98517.
3. Singh, R., & Sukapuram, R. (2023). Suchetana chakraborty, a survey of mobility-aware multi-access edge computing: Challenges, use cases and future directions. Ad Hoc Networks, 103044, ISSN 1570–8705, https://doi.org/10.1016/j.adhoc.2022.103044.
4. Thantharate, A., & Beard, C. (2023). ADAPTIVE6G: Adaptive resource management for network slicing architectures in current 5G and future 6G systems. J Netw Syst Manage, 31(9). https://doi.org/10.1007/s10922-022-09693-1.
5. da Silva, D. C., Batista Jr, J. O. R., de Sousa, M. A. F., Mostaço, G. M., Monteiro, C. D. C., Bressan, G., & Silveira, R. M. (2022). A novel approach to multi-provider network slice selector for 5G and future communication systems. Sensors, 22(16), 6066.
6. Liyanage, M., Porambage, P., Yi Ding, A., & Kalla, A. (2021). Driving forces for Multi-Access Edge Computing (MEC) IoT integration i5G. ICT Express, 7(2), ISSN 2405–9595. https://doi.org/10.1016/j.icte.2021.05.007.
7. Shakarami, L., Ghobaei-Arani, M., & Shahidinejad, A. (2020). A survey on the computation offloading approaches in mobile edge computing: A machine learning-based perspective. Computer Networks. https://doi.org/10.1016/j.comnet.2020.107496.
8. Chen, Q., Wang, Z., Su, Y., Fu, L., & Wei, Y. (2022). Educational 5G edge computing: Framework and experimental study. Electronics, 11(17), 2727.
9. Liu, J., Wu, Z., Shen, J., Liu, J., & Tu, X. Artificial Potential Field-Based Resource Allocation for Mobile Edge Computing. www.ijicic.org/ijicic-180505.pdf.
10. Srivastava, Amit, Sajjaf Jawaid, Rajesh Singh, Anita Gehlot, Shaik Vaseem Akram, Neeraj Priyadarshi, and Baseem Khan. "Imperative role of technology intervention and implementation for automation in the construction industry." Advances in Civil Engineering 2022 (2022), pp 1–19. https://downloads.hindawi.com/journals/ace/2022/6716987.pdf?_gl=1*7v7hi*_ga*MTIyOTIzMzI2MC4xNzA0NDQ5ODQ5*_ga_NF5QFMJT5V*MTcwNzM5MjAxNy4yLjAuMTcwNzM5MjAxNy42MC4wLjA.&_ga=2.42580069.1797774941.1707392017-1229233260.1704449849
11. Choudhury, S., Kolodziejski, J., Seskar, I., & Raychaudhuri, D. (2022, October). Experimental evaluation of virtual reality applications running on next-gen network scenarios with edge cloud assistance. In Proceedings of the 16th ACM Workshop on Wireless Network Testbeds, Experimental evaluation &CHaracterization (pp. 16–23).
12. Chen, S., Hu, J., Zhao, L., Zhao, R., Fang, J., Shi, Y., & Xu, H. (2023). Key technologies related to C-V2X applications. In Cellular Vehicle-to-Everything (C-V2X) (pp. 235–270). Springer, Singapore.

Index

Note: Page numbers in *italics* refer to figures and those in **bold** refer to tables.

Taylor & Francis Group
an **informa** business

Taylor & Francis eBooks

www.taylorfrancis.com

A single destination for eBooks from Taylor & Francis
with increased functionality and an improved user
experience to meet the needs of our customers.

90,000+ eBooks of award-winning academic content in
Humanities, Social Science, Science, Technology, Engineering,
and Medical written by a global network of editors and authors.

TAYLOR & FRANCIS EBOOKS OFFERS:

A streamlined
experience for
our library
customers

A single point
of discovery
for all of our
eBook content

Improved
search and
discovery of
content at both
book and
chapter level

REQUEST A FREE TRIAL

support@taylorfrancis.com

 Routledge
Taylor & Francis Group

 CRC Press
Taylor & Francis Group

Milton Keynes UK
Ingram Content Group UK Ltd.
UKHW031132141024
449569UK00006B/254